THE SHIP

By the same Author

TO ORGANISE THE AIR
The Evolution of Civil Aviation
and the Rôle of Sir Frederick Tymms
the Flying Civil Servant

AIRSHIP NAVIGATOR
One Man's Part in the British Airship Tragedy
1916-1930

IN MY ELEMENT
An Old Pilot's Tale

E.A. Johnston
OBE, FRAeS

THE SHIPMASTERS

Two Hundred and Fifty Years of Seafaring

George Mann *of* Maidstone

E.A. Johnston
OBE, FRAeS

THE SHIPMASTERS

First published 2000
by George Mann

ISBN 0 7041 0305 2

Printed and bound in Great Britain
and published by
George Mann Books, PO Box 22, Maidstone
in the English County of Kent

This Book
is dedicated to my beloved daughters

Rosalind and Caroline

through whose veins
the blood of all these splendid people runs

Contents

PART THREE — CLYDE SAIL AND THE BIRTH OF THE STEAM SHIP

PART FOUR — EAST OF SUEZ

Introduction

AT THE START OF THE TWENTIETH CENTURY MORE THAN half of the world's shipping tonnage was British-registered. Every year the ship-builders' yards on those great arteries of commerce, the Clyde, the Tyne, the Wear and the Mersey, built nearly two thirds of the world's shipping tonnage. Before the Second World War the London River comprised the world's most mighty port. All this has vanished.

When I was a child my Father took me on board his ship whenever she docked on the Tyne and showed me the Mariner's mysteries—how to splice a rope, how to box the compass, how to run safely up and down a ship's ladders. Great ocean-going ships loading and unloading lined both banks and, behind them, were the huge gantries and cranes of the builders' yards and the gaunt skeletons of ships under construction whilst the water was alive with small vessels servicing the river's business. All that has gone.

When I was a boy I sometimes stayed with my Grandmother at Wallasey, and my Uncle Jim, who was a shore-based Marine Engineer, used to take me to see over his company's ships in the Birkenhead docks. I can still visualise the vast engine rooms with their tiers of polished steel walkways linked by a network of ladders, and the gigantic, gleaming crankshafts and piston rods, and the cavernous black hell of the stokeholds. All this has disappeared as though it had never been.

As a youngster I sometimes sat on the beach at Southsea watching the busy deep-sea traffic passing through the Solent. It was a never-to-be-forgotten delight when one of the great Atlantic greyhounds—the old *Mauretania*, or the *Bremen* or the *Ile de France* or the *Normandie*—came regularly as clockwork sweeping out of Southampton Water, turning smartly to starboard round the Calshot

Spit Light Vessel and then hard a-port round the West Bramble buoy into the deepwater channel past Ryde and the No Man's Land Fort outbound via the Nab Tower for Cherbourg and New York. Years later I myself sailed across the Atlantic in both the *grandes dames* of them all, *Queen Mary*, and her sister *Queen Elizabeth*, little knowing that within a decade the Boeing 707 would have killed that truly luxurious traffic stone dead.

In my teens I stayed with Uncle Arthur and Aunt Kay who had the lease of the pier cottage at Mambeg on the Gareloch. Every morning promptly at 8am the elegant little paddle steamer *Lucy Ashton* chuffed alongside to collect Uncle Arthur on his way to his ship-broker's office in Glasgow. A couple of hours later, on her next scheduled round of the Gareloch's piers, Aunt Kay and Aunt Jess and my cousin John and I would board her for Garelochead for our morning game of golf. She was just one of a large fleet of steamers that plied their daily rounds of the Firth of Clyde and its lochs binding the shore communities together.

One day we took one of the steamers from Craigendoran to Glasgow's Broomielaw through an astonishing fifteen miles of busy shipbuilding yards and crowded docks. The paddle steamers, the shipping, the docks and the yards have all vanished . . .

The Gareloch then was the anchorage for something like thirty ocean-going passenger and cargo liners laid up as a result of the great depression, as were many other lochs abutting the Firth and up the west coast of Scotland. From time to time John and I rowed out to one or other of them to be greeted by the lonely watchman and invited aboard where we explored with flashlights the vast dark interiors of these ghost-ships. They were all pressed back into war service, most of them to be sent to a watery grave by German U-boats.

Throughout my working life, which was far removed from shipping and the sea, seafaring lay dormant in my blood though I knew that my father's family had been closely bound up with the industry. It was not until I had retired and was living in a cottage beside the Kirtle Water in Dumfriesshire that an idle impulse drove me to walk into the graveyard of a Kirk on a knoll overlooking the Solway Firth, starting me on a trail which has led to this book. To my complete surprise my eye fell upon a gravestone bearing the inscription, 'Frances, relict of Captain James Johnston of North Shields'. I knew nothing of the lady but did know that my great-grandfather of that name had lived in North Shields.

It was because I had fallen in love with this part of Scotland that I started to follow up the antecedents of Frances Livingstone, who, I found, had been born in a cottage a mile or so away from the one I had been living in. My research into the seafaring activities of my forebears, at first incidental, eventually took over; and once I had got into the pattern of it, it became inevitable that I would follow up my wife's maritime forebears as well.

The result, I hope, is not so much a family history as an account of how the courage and skill of a typical selection of fine men, supported by their remarkable women-folk, contributed to developing the British Merchant Marine which not so long ago dominated the Five Oceans and Seven Seas.

E.A. Johnston
Seaford, May 2000

So to the sea we came. The sea? that is
A world of waters heaped on hie,
Rolling like mountains in wild wildernesse,
Horrible, hideous, roaring with hoarse crie.
And is the sea, quoth Corydon, so fearfull?
Fearful much more (quoth he) than heart can fear:
Thousand wild beasts with deep mouthes gaping direfull
Therein stil wait poor passengers to teare.
Who life doth loath, and longs death to behold,
Before he die, alreadie dread with feare,
And yet would live with heart halfe stonie cold,
Let him to sea, and he shall see it there.
And yet as ghastly dreadful as it seems,
Bold men presuming life for gaine to sell,
Dare tempt that gulf, and in those wand'ring stremes
Seek waies unknown, waies leading down to hell.

Ed. Spencer—*Colin Clout's Come Home Again*

PART ONE

Sail on the Firth of Forth

1

Tradition

FROM TIME IMMEMORIAL THE 'PRINCELY FORTH', THE most navigable and commodious inlet of the sea on the East Coast, was the chief centre of Scottish shipping. The Isle of May and the Bass Rock loomed like guard-ships athwart the portals of the Firth. In spite of a stream running at six knots through the throat between the Ferry Peninsula on the Fife shore and South Queensferry on the Linlithgow shore, and swirling fiercely among craggy unlit islands—Inchgarvie, Cramond, Inchcolm, Inchmickery and Inchkeith—steep-to here, shoaling water there, there were always safe havens and places of shelter for ships in the waters of the Firth whatever the state of the tide or the behaviour of wind and weather. The harbours along the Fife shore were generally superior to those along the Lothian coast of the Forth, and of these Inverkeithing was the centre of a trade which furnished a considerable part of the Royal revenue until the middle of the sixteenth century.

Traditionally the Fife trades were salt, salt fish, coals and glass to England, the Netherlands, France and the Baltic, and northward and southward along the coast. Little came back from England, but Hamburg sent a lot of Rhenish wine and Hock, Norway traded great quantities of pitch and tar, deals and fir timber, Sweden sent copper, iron and timber, and from Dantzig, Konigsberg, Riga, Narva and St Petersburg came planking and oak and hemp and pitch and tar. From France, of course, the principal cargoes were wine and brandy.

As Inverkeithing decayed in the latter half of the sixteenth century, Burntisland and Kirkcaldy became ascendant for the next hundred years or so. By the time of the Union, however, Leith and Bo'ness had become Scotland's principal trading ports and for a while early in the seventeenth century the whole Baltic trade with the east of Scotland was concentrated at Leith. In the eighteenth century Bo'ness was second only to Leith for the French and Dutch trades. Aberdeen, Montrose and Dundee trailed far behind.

The Johnstons of old were bold and skilled seafarers on the eastern Scottish seaboard, mariners in the traditional Fife trades, who handed from father to son both the skills of intricate pilotage among foul banks, rocks and islands beset by swirling tidal streams in narrow waters, and the knowledge to enable them to enter difficult or exposed harbours in dangerous conditions of wind and sea.

Their lives were conditioned by the ebb and flow of the tides. The features of the bottom of the German Ocean and the Baltic Sea were etched upon their minds. In the thickest weather they could navigate their way around the northern coasts of Europe by the sounds and smells of land that penetrated the sea fog; and they knew to an inch the places of shelter in storm. Above all, generation after generation, they were inured to the loneliness of command at sea.

The Shipmaster was a worthy man in a business that could be profitable. At the top of the profession there were gentlefolk, as in England's East India Company, but in my family's circle it was the sons of prosperous minor merchants or of families with influence or connections in maritime circles who became mariners and passed on their wisdom from generation to generation. It was of such a line that the Johnstounes in Aberdour were sprung. The Masters bore responsibility for navigating and directing the sailing of the ships while the Mates supervised the doing of it. When it came to a fight with wind and sea, survival depended on the character and endurance of the Master more than any other factor. By contrast the seamen, those men who worked the ships, were recruited from the lowest ranks of society.

The Master was no perpetually supervised employee: he was the Owner's full representative, usually working beyond the range of direct control, answering only to the Owner (who generally wasn't there) and unto God; and as the Owner's full representative he acted independently on equal terms with merchants and factors ashore. Many

Masters had an interest in the profitability of the venture as part Owners themselves.

The most important factor in the advancement of a man to command—and in his staying there—was the recommendation of Owners. If he possessed the necessary personal qualities of courage, endurance and the will to dominate both men and the elements, and if he had mastered the essential skills of seamanship, pilotage and navigation, then the surest way to becoming a Shipmaster was to be the son of a Master of good reputation—such were the Johnstons.

Command was, nevertheless, always precarious. The status of Masters and their relationships with Owners varied enormously. Every contract was for a single voyage. In hard times there were always more Masters than ships. Even in good times a Master who was not able to live on air between voyages of a particular ship might find it necessary to seek another vessel as soon as he terminated a voyage. The good Masters, however, more often than not kept their commands for many successive voyages and sometimes for the rest of either their own or the vessel's working life.

Of old, the coastal and coal trades were a summer business and the ships were not often at sea for more than a few days at a time. Traditionally they were the nursery for training landsmen in the ways of the sea. The North Sea trades bred a tough seaman insensitive to his harsh environment, with a reputation for strength, courage and steadfastness afloat matched only by his capacity for wild drunkenness ashore. Through the generations my Johnstons not only had the lonely governance of these brutal men, but as youngsters serving their sea-apprenticeship spent their early years as dirty, neglected little ruffians messing for'ard with them.

It was a rough and punishing apprenticeship, yet the little brat hailing from a line of mariners could always be reasonably sure that it was the prelude to command.

That is not to say that it was difficult for any seaman who was sober and not altogether stupid, and who showed pride and skill in his trade, to advance through the ratings to quartermaster and boatswain and on to Second Mate or, in a coaster, even to Mate; but the key to command was not so much in having served his time as in having connections.

Although the true dividing line between the skills of seaman and mariner was the art of navigation, in the North Sea and coastal shipping the need for scientific navigation and mathematical calculations was minimal; the skill of the Johnstons and their likes was 'to know the land as soon as he shall descry it', and the Master's knowledge of soundings and of the faces of the coasts around northern Europe was an essential part of his fitness to command far outweighing skill with cross-staff, quadrant and almanac. It was only half in jest that men spoke of 'pilotage by the sound of dogs barking.'

In the 17th century the majority of the vessels engaged in the coastal trade were small ones of less than fifty tons burthen. The most popular among them were the two-masted ketches bearing square sails from their yards, and there was a profusion of smaller single-masted hoys, doggers and shallops. The Baltic, North Sea, and French wine trades were served in the main by ship-rigged vessels between fifty and three hundred tons burthen: that is to say, vessels having three masts—fore, main, and mizzen—each surmounted by a topmast which in turn carried a topgallant mast. Apart from the head-sails and the spanker or driver set on the mizzen-mast, all the sails bent on the yards were square. Although the sails of most ships carried bonnets in the 1650s, the practice of reefing rapidly superceded them by the end of the century.

The trade of the Fife ports had been virtually destroyed by the Cromwellian invasion of 1651 and the subsequent 'Usurpation'. After the Restoration, the English Navigation Laws restricting the carriage of practically all goods into or out of England and its overseas possessions to English ships manned by Englishmen, and which classified Scots people and ships as foreigners, not only killed the revival of the Scottish shipping trade with England but also largely denied employment in English ships to Scots seafarers for the next four decades.

Although stout efforts were made by the Forth shipowners to rebuild the Holland and Baltic trades they were stultified by the Anglo-

Dutch War of 1665-7. Burntisland never really recovered, and Kirkcaldy did so only briefly on the accession of Dutch William and Mary in 1688.

Aberdour Bay, finely sheltered and guarded by the ancient castle of the Earls of Morton on the cliffs above it, was historically the convenient landing place for people coming from the continent to the Scottish Royal Court at Dunfermline. From the high wooded bluffs to the east of the bay there was a splendid view across Mortimer's Deep to the old monastery buildings on Inchcolm to which the lands of Aberdour had belonged in ancient times; and beyond it, beyond Inchmickery and Cramond Isle, across the Firth to the Queen City of Scotland, her Castle Rock backed by the dancing peaks of the Pentlands and the dark range of the Lammermuirs along the south-eastern horizon.

Robert Johnstoune, who married Margaret Finlasone at Aberdour a few months before the restoration of King Charles II to the throne of Scotland, sired a dynasty of mariners who survived through the hard winter of smuggling from the time of the act of Union in 1707 until the aftermath of the Napoleonic Wars a century later, and thereafter prospered in lawful traffic on the seas.

Here as a child in the 1660s Robert Johnstoune's son James, when he was not playing in the ships' boats in the little harbour or on the old hulks lying in the dried-out sands at the bottom of the ebb, sat and watched the crowded sails on the gleaming water, identified the distant rocks of Fidra, Craigleith and the Bass and the perfect cone of North Berwick Law and dreamed of the future when he would sail his own ship past them and bear away to the southeast and the wide world beyond. Although Aberdour was possessed of a few small vessels engaged in trading within the confines of the Firth, its habitant seafarers found their employment mainly at Burntisland two and a half miles to the east or Inverkeithing not much further west, as well as Leith and Bo'ness on the Lothian shore.

Shortly after the English and the Dutch made peace in 1667 the ten-year old James Johnston went to sea as a cabin boy until he was old enough to begin his formal apprenticeship. As a lad with good connections he was able to profit from what few opportunities there were in the struggling coal and traditional trades out of the small ports of Fife. Slowly these increased as a result of the loosening of the monopolies of the old Royal Burghs by the 1672 Act of the Scottish Parliament, but when he finished his sea-apprenticeship and became a Mate in the early 1680s the times were out of joint and there was no scope for him to become a Master before his early death.

His wife bore him two sons, John and Edward, who managed eventually to make a living out of seafaring, but the dice were loaded against their father. English trade was protected by prohibitions or duties: the duty on salt hit the trade from the Forth severely; the French progressively stepped up duty on imported coal and then forbade the importation of salt herring and, later, wool; the skins exported from Scotland to the Baltic got fewer and fewer as more and more cattle were driven to England. For two years until the victory of the English Fleet over the French at La Hogue there was an almost total slump in the French and Baltic trades.

At the tail end of the Revolution against the settlement of the Scottish Throne on Dutch William and his wife Mary a small group of Jacobite Officers seized possession of the Bass Rock and held it until 1694 with the help of supplies and two large vessels obtained from the French. The Scottish Government finally forced its surrender after a long and rigorous blockade. James Johnston did his stint from time to time in vessels commandeered to maintain the blockade, including the frigate *Providence*. He served during the final weeks under a Master from North Queensferry, James Brown.

A couple of years later he was given the job of Mate to Brown's son John, to bring to Leith the *Instauration*, then being built at Hamburg, one of the ships of the new 'Company of Scotland Trading to Africa and the Indies'. Prepared at Leith for the expedition to set up a trading colony in Darien, she was renamed *St Andrew* and became the company's Expedition Flagship. James Johnston obtained the Mate's berth in her when the first Expedition sailed from Leith in July 1698 bearing the Imperial dreams of Scotland. But the Darien Expedition turned into a fatal disaster for the colony as well as a financial disaster for all of Scotland. The *St Andrew*, with James and all those hands who had survived the terrible sickness of the Darien spring, was lost at sea the following June after the abandonment of the first settlement.

By then his eldest son John was at the start of his apprenticeship in sail under the aegis of his uncle of the same name, Master of a small trading vessel out of Kirkcaldy. The closing years of the century provided the worst harvests and the worst and last famine in the history of Scottish agriculture, so the young John served the first part of his time sailing in search of grain from the southern Baltic ports. He also began to learn a new line of trade which was to stand him and his descendants in good stead, that of smuggling.

Once again, however, an English war in the Low Countries blasted the tender plant of trade between the Forth and Holland, this time irretrievably. For want of opportunity in the traditional trades young John did not finish his apprenticeship but took instead a berth before the mast in a privateering expedition against the Spaniards on the Pacific coasts of the Americas.

It was not successful.

He lost his ship, but survived and worked his way back by a devious route to a Scotland that had now become united with England.

2

The Smugglers

THE UNION OF SCOTLAND WITH ENGLAND IN 1707 brought some relief to the Forth to the extent the English Navigation Laws now protected Scots shipping, but it also brought upon Scottish economic life the damaging imposition of wholly new taxes. As a direct result, smuggling grew rapidly into a major and a patriotic industry.

Almost as soon as the new duties on imports into Scotland began to be levied, a fleet of Dutch vessels carrying contraband goods visited the Scottish coast, found it profitable, and continued to do so. The Scottish seafarers, confined at first simply to running contraband in small boats between the visitors and the shore, soon learned how to compete with them, especially those like young John Johnston and his father's relatives who had already been in the business of carrying to continental ports English wool which had been smuggled over the border.

It was not long before the youngster was acting as Mate to his Uncle John running small contraband cargoes directly from Holland and France. Even while the shipping on the eastern Scottish coast continued to decline into a desperate condition he became a skilful and resolute Shipmaster, kept going by his family connections in the 'fair trade' which came to dominate the life of the coastal communities.

In Aberdour John wooed and won for wife Katherine Cousen who bore him many daughters and, in 1717, one son George. Inevitably the lad, when he reached ten years of age, became ship's boy to his father before starting his formal apprenticeship a few years later. It was about this time that the Johnstons, working out of Inverkeithing harbour, first became involved with old Andrew Gibson the carrier in Dunfermline.

George became his father's Mate just as the harbour at Inverkeithing was beginning to show its short-lived recovery. The state of trade in 1738 which caused the Town Council to lengthen the pier to cope with it was neither entirely legitimate, nor unconnected with the fortunes of the Gibsons.

Old Andrew was a carrier of goods in a small way in Dunfermline before ever the weaving of table linen prospered there; his brother John was a Dunfermline merchant. He was born a handful of years after the Union, at the very nadir of Scotland's trade depression, when more and more people were being forced into illicit trading as the result of the penal rates of duty on exports and imports and the rigid trading controls and prohibitions. Contraband tea, tobacco, brandy and silks were run ashore all round the coasts of Scotland by mariners like the Johnstons while organisations grew up, some of them on a large scale, for distributing the contraband throughout the land.

This became the real Gibson métier. Their entrepreneurial skills and connections as carriers linked the smugglers on shore—the turbulent bands of men armed with blunderbusses, cutlasses, clubs and loaded whips who came to be known as lingtowmen in some regions—with those trading afloat, often some of the finest sailors and watermen in the world. Andrew Gibson's tentacles stretched out through the little port of Inverkeithing far beyond the confines of Fife.

In Scotland there was immense public support of illicit trading. What contributed to it more than anything else was the people's spontaneous indignation at what they perceived to be Parliament's tyrannical interference with their drinking habits by enforcing for the first time, in 1725, an impost on malt.

The predominant Scottish drink until the middle of the century was 'twopenny ale', so-called because that was what it cost to make the Scottish pint (equivalent to four English ones). The malt impost caused an immediate rapid decline of the brewing and drinking of 'twopenny'. Whisky, almost unknown outside of the Highlands before 1725, rapidly took its place. There was also a fast-growing demand for spirits smuggled from Holland and France. Scottish vessels plied a vigorous trade in contraband brandy between Rotterdam and the Forth, and the apprentice George Johnston was in the thick of it from the start. Vessels returning from Holland, France and Spain would rendezvous offshore with the colliers and coasters and transfer the high-duty goods as they passed up and down the coast. Similarly continental contraband-carrying luggers, particularly Frenchmen, quickly learned to hover close to the shore to do business with vessels in the home trade.

Gradually, however, the sheer weight of numbers of colliers out of the Tyne, ubiquitous along the coast from Orkney to The Downs, squeezed the Scottish vessels out. The Frenchmen came to exercise a protective paternalism over the Geordies when they were threatened by the Excise cutters, and eventually gave them a monopoly in the smuggling of brandy.

As Scottish vessels became increasingly unable to compete with the Tyneside ring, Andrew Gibson, then in his twenties in Dunfermline and finding his modest activities in the 'free trade' hampered by the decline of the Forth ports, decided to send his younger brother Robert to join the Scottish community in Sunderland with a view to organising the onward shipment of contraband to his Dunfermline connections. Why should he not take advantage of those tax laws which made it cheaper for English vessels to supply coal to ports north of the Tay than for the Scottish? But Robert did not find it easy to break into the institutionalised life of Wearmouth. The Tyne seemed to offer more scope. Although North Shields was then a poor, miserable place, a string of crowded wooden houses along the waterside below the high bank, it had the advantage of fronting onto a large, commodious harbour, the finest refuge—once its bar had been crossed safely—between Forth and Thames, and remote enough from Newcastle for the French to dare enter in search of customers for their brandy.

Its overcrowded squalor, the sheer density of the traffic flowing past its waterfront and its fierce independence of Newcastle, the centre of the sea coal trade, made it an ideal centre for nefarious activities such as the Gibsons were to engage in. Towards the end of the decade, having succeeded in making an opening for himself, Robert settled there and began to raise a family.

Between 1700 and 1800 the average tonnage of English coasting vessels doubled, and many a Scottish mariner found employment in them. George Johnston seems to have played some part in the Gibson family's ventures and to have prospered. Active and resolute in an era of depression, he relished the challenge of outwitting the myrmidons of the Englishmen's tax laws.

There was always a steady trade in lime between Sunderland and the Scottish east coast harbours and in coal from both Wear and Tyne; given a fair breeze the passage between Tyne and Leith Roads was a mere 24 hours sailing time. Using favourable winds and dark nights he and his like brought the art of running smuggled goods ashore to the peak of perfection. With the onset of the war of 1739-48, however, partly to obtain exemption from impressment into the Royal Navy, George enlisted Robert Gibson's help to take up employment in the London coal trade.

In Dunfermline a new prosperity was growing out of the weaving of Damask linen. During the decade following Robert Gibson's emigration from Fife the cottage weavers had flourished so much that they formed themselves into a Trade Guild and, in 1749, when his brandy trade was supreme and whisky was not yet a serious competitor, the British Linen Company was formed in Dunfermline. It was soon to dominate the town for generations.

The carriage of woven products to markets was a respectable cover for the illicit activities of the Gibson family headed by Andrew in Fife and Robert in Tynemouth, with whom George Johnston worked in his comings and goings to the harbour of Inverkeithing and the various offloading points between Saint Margaret's Hope and Kinghorn. His own many relatives along the Fife shore were also involved both on land and on sea. But at that time the 'free trade' in Scotland scarcely needed cover.

Smuggling there was carried out far more widely than in England, for in part it was a natural resistance movement in the face of duties imposed by an alien government in London, and as such it was deemed respectable by most social groups. Boats came to the green shores of Fife with perfect confidence not only of impunity, but also of a friendly welcome from the mass of the people. All ranks of the community shared in the trade's rich profits.

Robert Gibson died in the late 1750s having sired a sizeable family. Ann, his eldest child, married William Livingstone, another Scottish mariner sailing out of the Tyne, no doubt on nefarious Gibson business. Their great-granddaughter Frances Livingstone was to marry George Johnston's great-grandson James.

When old Andrew Gibson died in the early 1760s his son William assumed the mantle of head of family in Dunfermline. A younger son Andrew, who looked after the Culross end of the 'free trade', married in 1766 Margaret Brown of North Queensferry, a grand-daughter of the Captain James Brown who had brought the *St Andrew* from Hamburg to Leith with George Johnston's grandfather as Mate.

George Johnston did not follow only the London coal trade and smuggling in the Firth of Forth; he also made a number of trading voyages between London and British North America. In his forty second year he met a girl called Helen in Kirkcaldy, married her and settled down in the 'lang toun' to less of a roving life as Master of a lime sloop plying between Sunderland and the Fife harbours.

By this time Kirkcaldy had fallen to its nadir from its pre-Restoration place as Fife's chief port. It had enjoyed a vigorous foreign commerce, but by 1760 it employed no more than a handful of men in one 50 ton coaster and a couple of ferry boats each of 30 tons. The peace of 1763 at the end of the Seven Years' War, however, brought a rapid and substantial renewal of Kirkcaldy's shipping, checked only by the American Rebellion, such that thirty years later the port registry comprised 26 square-rigged vessels, one sloop and two ferry boats, employing about 225 men. Although some of the vessels traded to the Mediterranean, the West Indies and North America, most were engaged in the Holland and Baltic trades, and the smallest as coasters.

George never quite gave up the local profession of running contraband all along the Fife shore, at which he had become so adept in his youth. As a bold smuggler he was looked upon not only as a hero but as a patriot. He afloat and the Gibsons ashore were formidable allies in the constant battle against the Revenue men.

It was another son of Kirkcaldy, Adam Smith the economist, son of the Controller of Customs at the time of the Union, who described the smuggler as 'a person who, though no doubt highly blameable for violating the laws of his country, is frequently incapable of violating those of natural justice, and would have been in every respect an excellent citizen had not the laws of his country made that a crime which nature never meant to be so.'

After the end of the Seven Years War the chief watchdog against smuggling was the Revenue Cutter Service, built up from a force of about 22 ships covering the whole British coastline in 1763 to nearly twice as many in the next ten years. They were aided by a score of contract vessels. Many of the regular Revenue Cutters, commanded by fine seamen with high standards of service, did excellent work, but there were too few of them and their operations were not sustained by a sufficient intelligence service so as to become a sustained threat to the trade in which George was active.

When the American War of Independence broke out in 1775 the Revenue service afloat was cut back severely and later, there was substantial reform of the Customs Service and the use of Cavalry in support of the preventive Officers ashore. Even so, in the 1780s most of the tea and brandy imported into Scotland passed over the border into England as large gangs of Scottish smugglers continued to travel southwards from the myriad 'Brandy Holes' along the Firth of Forth.

3

Sea Coal

IN ANCIENT TIMES THE PORT OF ALLOA WAS ONCE pre-eminent, and for a long time was an independent Head Port with Admiralty jurisdiction from Stirling Bridge to Petticur on the north side of the Firth, and to Higgins Nook on the south side. The harbour was the best in the Firth above Granton despite its double tide at each flowing and ebbing—what they called a 'leaky tide'—and a range varying from 14 feet at neaps to 24 at springs.

The situation of the town was pleasing, nestling around the ancient tower that was the seat of the family of Mar, the mightiest coalmasters on the Forth. They were among the leaders of the expansion of the Scottish coalmining industry that followed the ending of the Seven Years War.

At Alloa, Thomas Lord Erskine trebled the output of his mines in the early 60s. The successor to the Mar estates, John Francis Erskine, to whom the title of 23rd Earl of Mar was eventually restored (it had been forfeited by his ancestor 'Bobbing John' Erskine for having led the rising of 1715), was also active in developing the collieries. The advent of Alexander Bald in 1774 as the manager of his mining enterprise saw the beginning of its most remarkable period of growth.

Although Alexander remained in charge of all the Erskine mining enterprises until he died in 1823, during his latter years his son Robert, the celebrated mining engineer, was probably the driving force in the technological advances which turned John Francis Erskine into one of Scotland's mightiest coalmasters.

At this time the shipping industry of Greenock, then the principal harbour in the Firth of Clyde, was severely set back as a direct consequence of the American War of Independence. In contrast to what had happened on the east coast, the Clyde generally and Greenock in particular had prospered greatly from the Act of Union, which for the first time had given to the Scottish people free commerce with the American colonies. The merchants of Greenock and Glasgow had been quick to seize the golden chance, but the huge fortunes they made in the tobacco trade were lost almost overnight when the colonial rebellion brought about its collapse.

James Harvey, shipmaster in Port Glasgow, was one of the mariners who sought to escape these consequences. He shrewdly took his family to Alloa to seek a living from its burgeoning trade in sea-coals. A number of factors combined to cause the substantial growth which began after the end of the Seven Years War: consumption of coal in the iron industry for one, notably at the Carron works; lime-burning, vital to the building industry which constructed the new town of Edinburgh from 1767; and a brief renaissance of the salt-boiling industry on both sides of the Firth.

James Harvey made a reasonable living in the coal trade out of Alloa until after the Peace of 1783 when, the North American and Caribbean cotton trade having exploded into profitable activity on the Clyde, he found it opportune to return there with his family.

He left behind his twelve-year-old son John, however, for the lad had started to serve his apprenticeship in a smack trading out of Alloa. He was serving as second Mate when the war against the French Revolutionaries broke out, and he avoided the risk of impressment by gravitating into the Government Transport Service.

When he married in 1807 at the age of 35 he was an established shipmaster. His bride was Charlotte Erskine, connected with the Erskines of Mar whose pedigree faded into the mists of antiquity long before the Norman conquerors brought their titles of feudal nobility into Scotland.

He established himself in Alloa as a respected shipmaster and shipowner in the ranks of the professional men who served the estate and the affairs of John Francis Erskine of Mar. His sister Isabelle married George Dick, a prominent merchant of Alloa and Edinburgh, and his eldest son John, who also became a shipmaster, married Jane, the daughter of John Bald the distiller of Carsebridge, a brother of Mar's colliery engineer and manager.

While the constant struggle of the coalmasters in Fife and Clackmannan to break into the London and foreign coal markets was throttled by the influence of the powerful traders of the Tyne and Wear, Scottish vessels were inhibited by differential taxes from trading north of the River Tay. There was, in consequence, a great deal of sea-carriage of coal within the confines of the Forth, and the entrepreneurial George Johnston was not too proud to make something out of it.

Edinburgh absorbed ship-born coal through Leith from the big collieries of Alloa, Fordell, Hallbeath and Wemyss across the Firth. At the head of the Firth of Forth Kincardine had a brief boom in the sea coal trade even though it lost to Alloa its position of importance in the general trade of the area. Alloa, given the status of an independent port in 1710, looked upon sea coal as its staple even though it became engaged in a wide range of trades. Inverkeithing handled the coal from Hallbeath. St David's, a mile or two east of Inverkeithing Bay, was developed by the Hendersons of Fordell who sent coals annually to London. Charlestown grew as an outlet for the neighbouring parish of Dunfermline.

To the east, Dysart, despite a tendency to silting, made substantial exports to Holland, as did Pittenweem and West Wemys. Kirkcaldy, though it had a 'tolerably good harbour but shallow water', was, like Methyl and Leven, of minor importance in the sea coal trade until after the Napoleonic Wars. George Johnston worked all these harbours.

During much of his early life as a shipmaster he would organise his own trade, sometimes lying off Leith for many weeks when sales were low, disposing of his cargo in small lots after long haggling. As the domestic market in Edinburgh grew, so merchants set up their coal yards near the harbour where they began to negotiate going freight rates with the shipmasters.

The practice of freighting whole ships spread to most of the other ports, and it was popular with shipmasters because it enabled them to achieve more frequent voyages and quicker turnarounds; the profit on a single voyage might be less but the overall gain was the greater.

George Johnston's wife Helen bore him several sons in Kirkcaldy, three of whom were to raise lines of shipmasters in North Queensferry and Kincardine-on-Forth. In 1765 she bore him a son who was christened James but came to be known as Hamish. With a clear eye to the future, George determined that all his sons should enter the sea-coal trade based on the Tyne and sought the influence of the Gibson connection to this end. And so it befell that the young Hamish, under Robert Gibson's protection, took a place at one of the several schools at North Shields where navigation was taught; and when he was ten years old he shipped on a trial voyage to London with Robert's brother-in-law, Will Livingstone.

From the second quarter of the 18th century there was a growing tendency to adopt two instead of three masts for medium-sized vessels. Before about 1720 the borderline size was about 50 tons, but by the 1760s it was about 160 tons. Up to the beginning of the Napoleonic wars the ship-rigged cat boats with a loading capacity of up to 600 tons dead-weight of coal dominated the seaways twixt Tyne and Thames; by the standards of the times they were proud, strong vessels; but there was no glamorous nonsense of figureheads or gingerbread decoration about the stern windows. They could be counted in hundreds taking shelter or awaiting a favourable tide. The crew of a 300 ton cat-boat in George's day was usually ten men and a couple of apprentices.

In the latter part of the century, the brig and the snow became the predominant types of craft. The difference between them was a slight one and they were both used for all general purposes, but the brig sailed mostly in coastal waters, while the snow sailed typically in the foreign trades. Towards the end of the century the coal trade and most of the trades with Spain, Germany, Holland and the Baltic were carried in brigs and snows.

The predominantly nor'easterly winds of winter not only put vessels at risk of embayment along the whole east coast, and most particularly amongst the sands of the Devil's Throat (as the Wash was known), but also built up hazardous conditions at the mouths of the northern English ports. For example, the channel at the Tyne Bar was only forty yards wide in those days. Until the advent of steam towage in the 1820s solved the problem of leaving and entering narrow harbours under adverse conditions of wind and tide, the tradition of laying the vessels up with yards and topmasts struck down through the winter months remained strong. At these times Masters as well as Mates and men sought to eke out an existence utilising alternative skills ashore, such as the cottage industry of handloom weaving along the Fife coast.

Hamish Johnston made his first voyage as a privileged boy, permitted to mess in the cabin, but although the passage was a fair one, such easy living did not keep seasickness at bay. Once the ship crossed the Tyne Bar she kicked up her heels on a fast broad reach, exhilarating sailing for the boy until black squalls coming out of Robin Hood's Bay set her lurching heavily and sent him retching at the lee quarter rail. There was little pleasure for him until the calm waters of Sea Reach above The Nore.

Will Livingstone took her close under Flamborough Head before laying a course direct across the Boston Deeps on a fetch for Yarmouth, where the collier fleet converged to crowd through the Cockle Gat. He brought her to anchor in Hollesley Bay. There she plunged deep into short, steep seas for four or five hours until the southgoing flood set in again.

When all hands were called to heave short on the anchor, Hamish was sent with the ship's boy into the cable locker to stow the filthy anchor cable as it came inboard—part of the process of hardening him against seasickness.

By now the wind had gone foul and the fleet of a couple of hundred odd sail beat through the narrow channels down the coast past Harwich and Orford, to lie in the Swin for the next tide, and so onward into the Sea Reach of the London River in a frantic race to be among leaders to reach the Pool in time to command the best prices and the quickest unloading for a rapid turnaround.

Although a legal system of turns for loading colliers had been instituted in the Tyne in the 1760s, in the London River there was no official regulation of shipping until the end of the century. Will Livingstone's owners bought the ship's load of coal, but he could not sell it on their behalf until he had delivered his ship's papers to the Customs in the Pool, arranged payment of the King's duties and the City's duties, and received a warrant authorising the Coal Meters to deliver the ship. Only then was the Master or the Owner's Crimp (who might be the broker, the undertaker or the wholesale dealer) in a position to negotiate a sale of the cargo at the Coal Exchange, which was then on the east side of Billingsgate Dock. Once a bargain was struck, Master or Crimp arranged for Meters, labourers and lighters to deal with the unloading of the vessel. Delivery might take three or four days or even longer.

Collier becalmed off the Pool

Will Livingstone knew that, with something like 200 vessels racing for the Pool, even if he were among the leaders there might be days of waiting before he could start to unload, but if he were a tail-ender he might wait for weeks. A quick turn-round depended largely on a Master's ability to get early clearance' from Customs and the City Cocket Office as well as on the Owner's willingness to pay 'despatch money' through the Master.

Some Masters sent a man, or more usually a boy, away from the ship at Yarmouth to race overland with the ship's papers to the Custom House in London. Livingstone, however, reluctant like many to lose useful hands for the most exacting part of the passage, did not send the ship's boy away until reaching the legal boundary of the Port of London at Gravesend. Hamish went with him. Their instructions were to make for the Custom House, but if the Clerks were not in attendance (their hours were from 10 am to 3 pm) the boys were to make for a certain public house where the cockets could be entered at any hour into a book kept for the purpose of regulating the ship's turn for the Coal Meters.

Whilst waiting for their vessels to enter the Pool, a large number of ships' boys were thus thrown together in public houses and other places of low resort; but Hamish had been directed to go on to Wapping to seek one of the Gibson clan who kept an Inn there, which was where he stayed until ready to sail back to Shields. He did not like London.

Shortly afterwards he made another two trial voyages. He did not mess in the cabin as before but was made to shift as other boys did, dirty from performing all the most menial jobs on board, kicked, spurned and neglected. Having none the less confirmed his determination to follow a mariner's life, Hamish was formally apprenticed in 1779 for seven years to Will Livingstone, still the Master of a cat-boat of about 350 tons burden, a good sea boat but very old. For this service he was paid a total of £25.

After carrying coals to London the vessel was laid up for the winter and Hamish was sent to the Trinity School at Newcastle to improve his navigation.

Early in spring, France went to war with England in support of the American colonists and there was an immediate flurry of impressment into the Navy, with the result that by the year's end Hamish's ship was being worked by nearly a dozen apprentices out of a total of twenty hands to make up for the shortage of trained seamen.

4

War and Peace

HAMISH JOHNSTON MADE HIS FIRST SUBSTANTIAL BALTIC voyage in 1781 when he was in his second year as an apprentice under Will Livingstone. By that time Holland, as well as France and Spain, was in the war, and the convoy system was in force.

Having loaded coals for Luebeck, Livingstone sailed up to Leith Roads to lie for several weeks awaiting the arrival of other vessels for the convoy. With his native Fife shore beckoning him across the Forth day after day Hamish might have been homesick had he not at that juncture been appointed to the position of Cabin Boy, a most onerous post at once steward, butler, waiter, shoeblack and char, where kicks and curses from a tyrannical Master were the order of the day.

After ten weeks they sailed for a pleasant crossing of the German Ocean under the escort of a naval squadron which deserted them to seek out and engage the Dutch fleet in the Dogger Bank action of August 5th. Livingstone left the convoy at Elsinore to proceed alone to Luebeck where he anchored in shoaling water and transferred the cargo to bilanders. After that he sailed in ballast to Cronstadt to load a cargo of hemp at the mole for Hull.

The season was late, the Baltic and the German Ocean both stormy and the return passage in the teeth of the prevailing winds was long-drawn. They spent some time lying off the neutral port of Elsinore waiting for convoy to England, but once clear of the Kattegat the convoy was dispersed by stormy weather and some vessels, including Hamish's, had to seek shelter for a while in Norway until the wind came from the northeast. It was full winter when they arrived back in Shields after a nine-months' absence, and unrove the running rigging and sent down topmasts and laid their vessel up until spring.

By then peace had been declared, and Hamish was in the crew on an independent passage in ballast to the Finnish port of Wyborg where a cargo of deals was got in smartly, both in the hold and on deck. The vessel beat back to Newcastle without incident to deliver the deals and take on coals for Cronstadt. There she loaded a freight of iron, hemp and tallow.

Although in this second half of the year the return passage to Newcastle was featured by gales again, the whole voyage lasted but three months compared with the single nine month wartime voyage of the previous year, and two were made this season.

Early next spring, after the winter lay-up, Hamish made several trips in the coal trade to the London River before setting out in ballast at the beginning of June to pick up a freight of tar from Archangel. Fair winds bore him briskly to the snow-covered North Cape. In the White Sea it was already hot; in Archangel mosquitoes plagued him. Before the tar could be loaded from flat-bottomed praams, each carrying 1,000 barrels, the ceiling of the hold had to be caulked and paid over with clay.

The return passage in quickly shortening days and contrary gales was uncomfortable and cold. Abreast the Firth of Forth particularly severe and prolonged gales drove them to seek shelter in St Margaret's Hope for several days.

Somewhat privileged as senior apprentice, Hamish was permitted to go ashore to visit Robert Gibson's cousin Andrew and his wife Margaret Brown in Inverkeithing.

The port which had been in ruin in 1775 had acquired a new importance in the 1780s, the removal of the discriminatory taxes on the coastwise movement of Scottish coal having resulted in the beginning of a boom of the Fife coalmining industry that was to last for a couple of decades.

The construction of the waggonways from Hallbeath and other collieries in the neighbourhood of Dunfermline to Inverkeithing Bay created an annual shipment from that port of about 25,000 tons, and sometimes as many as 40 or 50 sail lay off the bay awaiting loading.

This run ashore rekindled Hamish's interest in his native coast. His father had been dead these five years but Andrew Gibson confirmed that Inverkeithing held opportunities for an able young mariner in the traditional trades with Holland, the Baltic and Scandinavia as well as participation in the coastal trade at those times of year when ice blocked the Baltic.

Discreetly Andrew spoke to Hamish's Master and satisfied himself that the youngster showed high promise. So it was arranged that the following year he sailed under Andrew's brother-in-law George Brown of North Queensferry as Master.

In the very close community of the North Ferry, the Browns were a large maritime dynasty stretching back to the previous century. Within a couple of years Hamish married his skipper's daughter Elizabeth and set up his home there.

North Queensferry was peopled by the tiny, close-knit community of ferrymen of the Queen's Passage, so named after the saintly Margaret, consort of Malcolm Canmore. For nearly 800 years these ferrymen had claimed hereditary rights to operate the ferry. They guarded their monopoly jealously. The men did not admit strangers. No man born beyond the Ferry Hill was allowed to serve in the boats. But the rewards were not high and the more enterprising young men forsook the ferries for the Navy or the Merchant fleets.

The Browns, very much part of the ferry community in origin, were just such enterprising men. George Brown was born in 1738, the son of a North Ferry shipmaster William Brown and his wife Margaret Steel. Near the gateway at the site of the ancient chapel there is a small graveyard with the inscription 'This is done by the sailers in North Ferrie 1752', and upon a recumbent stone bearing the initials WB and MS—not improbably William Brown and his wife Margaret Steele—is the delightful epitaph:

'Heir we lay at anker with manie
in ouer fleet in hopes to way at
the last day ouer Admiral
Christ to meet.'

At the Queen's Ferry the waters of the Forth suddenly narrow to little more than a mile, with the island of Inchgarvie standing like a huge stepping stone two thirds of the way across. Here the northern shore looks invitingly close and the waters less menacing than the open seven miles or so that lie between Leith and the Fife coast, but the squalls off the high cliffs and the fierce, swirling currents that flow through the narrow, curved funnel rendered the passage treacherous for small vessels.

For a short period about 1788 the owners of the Ferry Passage, driven to distraction by the way the hereditary ferrymen abused their monopoly, let out the operation of the boats to a group of ex-Naval seamen from the Tyne, one of whom, Robert Nelson, became friendly with Hamish Johnston. This operation, however, was a short-lived disaster, for the foreigners lacked the essential experience and local knowledge to navigate such treacherous waters safely at all times.

During 1794 Hamish was engaged in transporting the Scottish regiments from Fort George to Flanders, and in the following year his wife bore him a son William at North Ferry.

In 1797, when the whole coast of Europe from the Netherlands to Northern Spain was in the power of France, and the French, Dutch and Spanish fleets were all ranged against the Royal Navy, and a French attempt at an invasion of Ireland had been defeated not by the Navy but by adverse winds and gales in Bantry Bay, Hamish and his Brown in-laws signed the following petition addressed to the Lord Provost of Edinburgh, who forwarded it to the Commander-in-Chief of the Forces in Scotland:-

> "We his Majesty's loyal subjects, inhabitants and seamen of North Queensferry being informed that our enemies threaten an actual invasion of Great Britain, and it is not certain where the attempt will be made, do make a tender of our services in the Defence of our King and Country and the British Constitution at the Battery of Inch Garvie, North Ferrie and Island Colm, if required, signed this third day of March 1797:
>
> "James Brown, Edward Brown, John Greig, William Brown, Robert Malcolm, John Ellis, Alexander Graham, Alexander Ritchie,

David Bartholomew, Robert Main, John Burn, David Smith, William Anderson, James Marshall, William Fyfe, Charles Thomson, Collin Read, George Jeameson, Robert Smeaton, John Inglis, Thomas Malcolm, Peter Millions, George Brown, Alexander Jamison, Archd Hinderson, William Brown, James Johnston, William Forrester Snr., John Forrester, William Forrester Juner., Alexander Siegge, Thomas Ogilvie."

Leith was at this time rapidly growing in importance as a rendezvous for the assembly of convoys. About halfway between the Queen's Ferry and Alloa, on the south side of the Firth within the jurisdiction of the head Port of Bo'ness, stood the vast Carron Iron Foundry, then the most extensive in all Europe. It was a vital source of armaments: cannon, mortars, carronades, shot and shell as well as smith-work and machinery of all kinds. A small fleet of transports, mainly former collier brigs, was constantly in service to carry the Carron products to the naval dockyards at Chatham, Deptford and Portsmouth and Hamish easily found employment in this, the Government Transport Service.

Inevitably miscalled Amos in documents compiled by English shipping clerks , he eventually became Master of the brig *Hazard* of Leith in this occupation which employed many of the sounder mariners of the Forth ports.

One other such was John Harvey, who was serving as second Mate in a sloop-rigged smack out of Alloa when the war against the French Revolutionaries broke out. This was the principal vessel for carrying non-bulky freight in coastal waters.

Amongst other things it carried salmon and other fresh fish in a rapid transit to London. Out of this express trade there evolved a fine type of fast, heavily rigged and generously canvassed passenger-carrying vessel of 70 to 80 feet overall, about 150 to 200 tons burden One such, around 1820, was John Harvey's sloop *John and James* registered in Alloa, a prize vessel taken in the wars.

For several decades until the 1830s these sloops dominated the passenger trade between Leith and London. They were heavily manned and well maintained. Despite their small size (they provided two cabins, one for men and one for women) they had a reputation for comfort and reliability. The average passage took five days, but the record was supposed to have been a magnificent dash of 42 hours.

It is not unlikely that many a time the vessels of Harvey and Johnston lay together in Leith Roads awaiting convoy. They may even have been acquainted with each other. Harvey's sons, however, were to follow their fortunes on the Clyde whence the family had come, whilst Johnston's descendants sought theirs on the Tyne, and it was not until nearly one hundred and fifty years later that the two families became united in marriage.

Hamish Johnston's son William was born at North Ferry in 1795. With all the salty blood of his forefathers in his veins he was more familiar as a child with the bottom-boards of a coble than with a crib, and could scull almost before he could walk. After a few short voyages as ship's boy aboard the vessel of one of his relatives he was formally apprenticed in 1807. Britain's command of the seas was by then unchallenged. William made a number of passages to Heligoland, which had become the principal entrepot for smuggling British goods into the continent following Napoleon's edict barring all European ports to British trade. He sailed in one of the transports supporting the Walcheren Expedition in 1809, one of the largest amphibious operations that had ever been known. He sailed to Spain in 1811 and 1812 carrying stores and reinforcements in support of the military campaign there.

In between operations he was, like his father, regularly engaged in voyages to the Baltic, the vital source not only of naval stores—masts, tallow, hemp, and flax—but also of the grain needed to support the series of disastrous harvests in Britain.

After the final defeat of Napoleon in 1815, British ports were overrun by discharged naval seamen for whom there was no possible employment because of the catastrophic decline in foreign trade that followed. Hamish found a permanent berth ashore as ship's husband to his brother Andrew, older by a year, in business as a shipowner in a small way at Kincardine-on-Forth. Andrew's twin sons Andrew and Peter eventually became ocean-going Masters.

Within a few years of settling at Kincardine Hamish himself owned a small coaster which was skippered by his second son James, while his youngest son Robert became a shipmaster at North Ferry. Hamish's younger brother Robert had also gone to sea, and both Robert's son George and his grandson Robert became shipmasters.

Hamish's son William, second Mate in a transport, was to some extent sheltered from the immediate post-war recession. Having taken part in the desperate emergency transportation of Wellington's troops to Belgium in advance of the Waterloo campaign he was lucky enough to be in the position to continue in the service of supplying the army of occupation for most of the next five years. In the winter of 1815 his ship put into the Tyne to load a cargo of coals for Antwerp, giving him an opportunity to look up some of the Gibson connections. There he met an old acquaintance of his father's, Robert Nelson, who had been one of the Tyne sailors employed to operate the Queen's Ferry twenty-seven years before.

Nelson had a daughter Helen and she and William fell in love. He forsook Fife forever to marry her in Tynemouth in 1816.

5

Deep Water

AT ALLOA IN 1815 FRANCIS ERSKINE HARVEY WAS BORN to John and Charlotte. Like his father and his elder brother John, he served his apprenticeship sailing out of Alloa, which still had a considerable foreign trade. His time completed with great credit to himself, he sought a career in deep waters, and took a berth as second Mate in the full-rigged ship *Alfred* (342 tons) registered in Alloa.

Built in Hull in 1801 she had the old arched transom with heavy windows to light the great cabin. She carried pig-iron, woollen goods, glass and whisky to North America, and brought back timber, battens, deals and hemp.

By this time, however, Alloa had fallen far behind the rapidly growing port of Leith which, with Rennie's extensive harbour works, had become the pre-eminent port of the Forth. Most of the Baltic trade had been drawn off to Kirkcaldy on the opposite shore, but Leith was the centre of the considerable foreign trade with the rest of northern Europe, the Mediterranean, America, the East and West Indies, Australia and China. Following his voyage in *Alfred*, therefore, Frank Harvey sought a vessel sailing out of Leith, and for the next three years voyaged on the North Atlantic in the 312 ton brig *Prince George*, carrying variously linens, cottons, silks, woollens, haberdashery, iron hardware, machinery, coals and fish outbound.

She was a venerable vessel owned by Duncannon of Alloa, built in London in 1789 and not unlike Captain Cook's *Endeavour*. Before the American challenge came in mid-century, the European tendency had always been to develop merchantmen like long, deep, beamy boxes with rotund bows that shoved the sea before them, and comfortable afterbodies to carry the great stern-windowed cabin out of the reach of breaking seas. *Prince George* was representative of the small merchantmen of her day, strongly rigged, stoutly built, of moderate draft, and thoroughly seaworthy. She was neither spacious nor comfortable and certainly was not fast. She might not be handy, but lying to under reefed mainsail she could withstand the stormiest seas in the world.

She carried six square sails—a course, a topsail and a topgallantsail on both fore and mainmast—and five fore-and-aft sails consisting of a jib and staysail for'ard, a main topgallant staysail and main topsail staysail between the masts, and a spanker or driver aft. She probably made about seven knots on her best point of sailing with the wind a point or two abaft the beam. Close-hauled in a topsail gale, as when making the offing to weather Cape Finisterre before picking up the northeast trade winds, she probably sailed between five and six knots.

Because of the prevalent adverse winds and the rough weather in mid Atlantic, Masters of westbound vessels rarely attempted the direct route across the Western Ocean. The shortest distance between two points of an ocean passage was of little significance, for what a Master wanted was the quickest passage dictated by wind and weather. Westbound from the Forth to North America there were two practical alternatives: either the ancient Viking passage using the easterly winds on the north side of the cyclonic systems of the temperate latitudes, or the southern route discovered by Columbus.

Despite heavier weather the northern route usually provided the fastest passages to the Gulf of St Lawrence via Belle Isle Strait, for the currents sweeping down from the Arctic added their help to the generally favourable winds; but it was only practicable in late summer and autumn when the ice had receded and the Newfoundland Banks were free of ice.

Frank Harvey sailed this passage from Leith in the latter half of 1837, first working north about between the Orkneys and Fair Isle. On her subsequent voyages *Prince George* bound from Leith to Canadian ports usually made her way down the German Ocean and then faced a long slog to windward down Channel, leaning hard on the southwesterly gale that funnelled from the Atlantic through the narrow seas, until she had sufficient offing to avoid being set into the Bay of Biscay.

This early part of her voyage seeking the fair winds and smiling seas of the southern route to the Americas was beset by complicated tidal streams and lay between dangerous, iron-bound coastlines where a square-rigged vessel had nothing in her favour save her own seaworthiness, the stout heart of her Master, and a Mate possessing the skills and qualities of character capable of translating the Master's intentions into performance.

Thus *Prince George*, having made her offing, set course to pass within sight of Madeira and into the northeast trade winds in about 30° north latitude, whence she wore away to run westward until in 50° west longitude. There she edged off to the northwest, passing a couple of hundred miles east of Bermuda and pointing direct for Halifax or Cabot Strait as may be, allowing the set of the Gulf Stream coming up on her port quarter to help her.

This was indeed working the great swirl of winds and water of the oceanic system that Columbus had discovered, sailing westward on the northeasterlies and then drifting north and east on the Gulf Stream into the zone of the westerlies and the variables.

The Portuguese had known the northeast trades long before Columbus, but it was not until Prince Henry the Navigator opened his school at Sagres over 100 years later that the complex nature of the trade wind system began to be understood through the correlation of observations made on many a painful voyage. It was his discovery that the quickest passage from Western Europe to the Cape of Good Hope was not to try and crawl along the African coast as all European sailors were accustomed, but to stand boldly out to sea keeping first the northeast and then the southeast trades square on the beam.

Later it was the Dutch who learned how to use the northeast and southeast trades to sail down into the permanent westerlies of the Southern Ocean and so make fast passages to Indonesia. Through these discoveries the mariners of Europe enabled merchants to profit from the rich trades of the New World.

In 1840 Frank Harvey joined the 400 ton barque *William Mitchell* of Alloa as Mate for a voyage to Australia. The eponymous owner was a large general merchant in Alloa. One of his sons was Master and Harvey no doubt had some interest in her or her freight. She sailed from London in March, but before she arrived back in home waters early in the following year her Master had died and Harvey had taken command. At the age of twenty-six he had acquitted himself well enough that the older Mitchell, a long-time acquaintance of his father, confirmed his appointment for the next and longer voyage.

In later years the clippers seeking rapid passages for their specialist cargoes—tea from China, wool from Australia—made non-stop voyages from port of origin to destination, subordinating everything to speed. Frank Harvey in *William Mitchell*, however, was more representative of the great fleets of general traders which wandered world-wide over the face of the waters carrying cargoes from port to port. When *William Mitchell* cleared Leith on July 19th her crew was not to see the place again for eighteen months. Her immediate destination was Rio de Janeiro.

Sailing north about, she lay at anchor in Long Hope in the Orkneys for the best part of a week awaiting a fair wind so as to be able to make the customary offing in about 12° west. A Master's three cardinal rules of ocean passage-making were simple enough: never to hug the wind when it was foul, but to let the ship go at least one point free through the water; always to profit as much as possible by well-known prevailing winds and currents; and with a foul wind, to lay the ship's head on the tack that will enable her to look best up for her port.

The whole art of passage-making was to keep going, making progress somewhere in the right direction. And so *William Mitchell* progressed under Frank Harvey's prudent guidance; having passed within sight of Madeira into the northeast trade wind she was in 14° north, 26° west, well southwest of the Cape Verde Islands, on the thirtieth day after leaving Long Hope. Once in the tradewinds the ship took on an almost idyllic life between a sea and a sky of deepest blue, booming steadily along through the warm air in a peacefulness broken only by the creak of the wheel, the hiss of the sea at the cutwater and the tolling of the hours on the ship's bell. The watches were reduced to only three or four men: one to steer, one for the lookout and one on standby, for the ship virtually worked herself through the tranquil weather once the light sails had been bent onto the yards.

Harvey's intention after leaving Cape Verde behind was to keep to the meridian of 26° west so as to minimise the length of passage through the doldrums. At that time of year he expected to loose the northeast trades in about 11° north latitude and pick up the southeast trades in about 5° north, but with luck he should ease his passage through the doldrums by catching the westerly skirt of the West African monsoon.

Sailing through the doldrums was hard work, for the heavy yards had to be hauled round to catch every catspaw; without wind to help, they were a dead weight. The courses had to be hauled up in the calms to prevent chafe as the ship rolled, then set again to catch each zephyr. As southwesterly airs predominated, Harvey tried to keep the ship on starboard tack, for he was anxious to cross latitude 5° north at about 18° west so as to have just that extra bit of sea-room to weather Cape San Roque on the eastern tip of Brazil.

The calm was punctuated by a succession of heavy rain squalls, and he ordered the big water-tanks on deck to be filled from buckets and barrels.

While the salt spray was not flying, all hands took the opportunity to wash their bodies, their clothes and their blankets. Here, too, was the opportunity for some shark fishing—not so much for sport as from a deep-seated urge to kill as many of the hated beasts as possible. Customarily the first to be killed had its tail nailed to the end of the bowsprit to divide the headwinds. The backbone was good for a walking stick.

Harvey's transit of the calms and thunderstorms and the oily sea in the doldrums had been a copybook exercise in how to do it efficiently in the minimum time, and after a few days the wind set in firmly from the southeast. He wore the ship onto port tack, and with the yards hauled onto the backstays *William Mitchell* sizzled along closehauled on the trade wind, scattering clouds of flying fish before her. On 17th

September she was on the equator in longitude 21° west, standing across the wind in the knowledge that as she advanced it would draw further east and enable her to lay a course to weather Cape Frio.

Early on 30th September she arrived off the entrance to the harbour at Rio, having made a better than average passage from Long Hope. It had as usual been superb sailing in the tradewind zone under balmy, cloud-puffed blue skies, but there was nothing hit or miss about it. The principal sailing routes and their seasonal variations were well-established and documented; a literate and studious Master had only to correlate the recorded data with his geographical position and his observations of actual wind and weather to be able to maintain a reasonably average passage at any time of the year between the ports that ordinarily offered trade.

In the afternoon, having lain offshore until the sea-breeze was established, so as to avoid the risk of running into a calm and being thrust by the heavy swell and the currents onto one or other of the many rocky islands at the mouth of the bay, Harvey sailed through the magnificent portals formed by Sugarloaf Mountain and Santa Cruz into the roadstead of Rio de Janeiro, and dropped anchor in four fathoms of water off the city, a cable's length northwest of the Isle of Cobras. When he went ashore, however, he discovered that the facilities were far short of the potential of the harbour.

The quay was inadequate and there were no docks, and the facilities for unloading his bales of woollens and linens were indifferent. In precisely a week, nevertheless, cargo was discharged and a load of coffee taken on for Melbourne. On 6th October *William Mitchell* put to sea again with ten thousand miles of open water in front of her before ever the crew would espy land again.

Rio at that time of year is on the edge of the variables, and a fresh northeast wind is not uncommon off Cape Frio. As Harvey had calculated, so it befell, and he was able to lay a course standing off to the southeast through the belt of variables until he began to pick up the northwesterlies. The flying fish grew fewer, and then an albatross appeared, and a couple of stormy petrels, the first birds they had seen since entering the tropics. These heralded cold and fierce winds.

Harvey ordered the heavy weather sails to be bent on the yards. He passed Tristan da Cunha and Gough Island without seeing them. When he reached latitude 40° south on the Greenwich meridian he laid his course due east. For the next 5,000 miles, on a dead run with the yards squared to the wild west wind, the threat of icebergs was constantly on his mind. Great grey seas rushing up from astern crashed past the ship under frowning grey skies.

In later years the clippers, seeking speed at all costs, would take an even more southerly route, but the general run of traders preferred the comparative comfort and security of the smoother seas north of forty south where the winds were seldom so strong as to prohibit carrying sail.

Harvey saw nothing to be gained by going further south into frequently and conspicuously more stormy weather with terrific seas and violent shifts of wind. As things went, he raised Cape Otway on 15th December, 72 days out of Rio, a somewhat longer than average passage which he ascribed less to sticking to the fortieth parallel than to a rather slow passage through the variables between Rio and 35° south.

William Mitchell's anchor went down in five fathoms off Williamstown in Port Phillip Harbour on December 16th and there, after discharging, she lay forlornly until early autumn. Her next port of call was Singapore, sailing in ballast to pick up a freight there, but Harvey decided against a premature departure while the northwest monsoon was dominating the seas north of Australia and giving promise of a long, tedious passage against contrary winds.

On the east coast of Australia in January and February the winds, though variable, tend to be northeasterly and conspire with the southgoing coastal current to delay northbound mariners. If on the other hand a Master was tempted by the frequent easterly winds to the south of Australia to sail west about for Singapore at this time of year, he was likely to be trapped by the prevailing northerly winds and southgoing currents beyond the chain of Indonesian Islands: it was not unknown for a vessel to take upwards of thirty days for the five hundred mile passage to windward between Sunda Strait and Singapore.

Harvey, having talked these considerations over with his fellow Masters on many an evening, wisely deferred his departure with a view to reaching the entrance to Torres Strait early in April so as to miss the worst of the equinoctial gales and to be reasonably sure that the favourable southeast monsoon was established in the Arafura Sea and beyond.

With fair winds all the way he expected to make a quick passage to Singapore taking the route outside the Great Barrier Reef to Bligh Entrance, thence through the Torres Strait and out by the West Entrance, across the Arafura Sea, avoiding its extensive coral shoal areas by going through the Wetar Strait north of Timor into the Flores Sea. Fair winds and good weather held as he sailed across the Java Sea, through the Karamata Strait and Riow Strait into Singapore, arriving on 26th May, forty-five days out from Port Phillip.

It was, by and large, an adventurous route for him to take through intricate waters heavily infested with coral. It would have been a less continually anxious passage, perhaps, had he caught favourable winds in the Australian Bight and gone south about and up into the western Java Sea through Sunda Strait. That he chose his first passage through coral by way of Torres Strait is indicative of the character of the man, for he would have read the remarks of Matthew Flinders who charted it in 1791/92: 'Perhaps no space of three and a half degrees in length presents more dangers.'

He found that accurate pilotage, constant attention to the freakish currents among the islands and shoals, and above all keen and intelligent lookout were essential if disaster was to be avoided.

The most dangerous characteristic of coral formations is that, being almost vertical in development, they usually have deep water close to. Soundings therefore rarely indicate the proximity of reefs. Although there are several channels through Torres Strait, very little of the area was properly surveyed or charted. Indeed, twenty years later the full-rigged ship *Severn*, the Mate of which was to become the father-in-law of one of Harvey's granddaughters, struck and for a whole day was stranded on a coral reef at the entrance of Bligh's Passage.

Leaving Singapore for Antwerp on 29th June Harvey was faced

with a long, tedious passage through the five hundred miles of doldrums to Sunda Strait. It took him thirty days, ghosting east out of Singapore and working south by Karimata Strait along the west coast of Borneo. The wind was light as well as contrary.

But having cleared Sunda Strait on July 29th he picked up the fair winds of the southeast trades and there was nothing but fast sailing for the next 5,000 miles across the Indian Ocean, south of Rodriguez Island and Madagascar until, meeting the variables off the south coast of Africa, he picked up the Agulhas Current which he used to help him beat against the mainly westerly variables round the Cape of Good Hope into the South Atlantic.

Once past The Cape and with sufficient offing to the northwest *William Mitchell*, again in the trade winds, tramped down the miles to St Helena where on 5th October Harvey anchored and sent a boat ashore to pick up fresh fruit and vegetables.

From St Helena he steered to leave Ascension Island close to starboard; then, expecting to lose the southeast trade just north of the equator, he aimed to cross it in about 28° west where he thought the belt of the doldrums would be at its narrowest for the time of the year.

All the way up the Atlantic Harvey kept the crew at chipping and scraping and painting. As soon as he lost the southeast wind he made the best northerly course possible to reach the northeast trade. Having caught it and found it brisk enough to stow royals, he held *William Mitchell* a point free of the wind on starboard tack.

She dove fast, heeling well over with heavy sprays sweeping over her weather rails and filling the lee scuppers. When the wind died in 28° north, 40° west, the ship lay for several days wallowing in calms and fickle winds until suddenly, out of a squall, she picked up the westerly system.

Harvey shaped course to pass fifty miles west of Flores, thence to the English Channel. The sea changed from tropic blue to deep green.

For a Mariner entering the Channel from westward the soundings are a poor guide because a similar sounding and bottom may be found at places a hundred miles apart. A more useful guide was to take a line of soundings carefully and lay them down. Harvey took the preferred course of approaching Scilly and fetching the Lizard rather than making his first landfall in the vicinity of Ushant where the strong tidal inset added to the hazards of the rocky coast.

Near Scilly 60 fathoms is a safe depth, and off the Lizard 49 fathoms, and between the Lizard and Start Point 40 fathoms. Approaching Land's End the chief set of the tides and currents is to the northward, and in thick weather Harvey found it anxious work for even his line of soundings did not seem to be a fully trustworthy guide to where he was. Only the Lizard was lit at night to help. It would be another 15 years before the Bishop Rock light was established on the Scillies. The Wolf Rock, half way between the Scillies and the Lizard and submerged to a depth of six feet at high water, lay unlit until 1869.

William Mitchell was in the Downs early on 24th November, one hundred and forty-eight days out of Singapore. That same evening she picked up a pilot off Flushing for Antwerp. She berthed there on the 29th, unloaded, and sailed in ballast for Leith Roads where she received orders for the Clyde. On January 16th 1843 she dropped anchor at the Tail o' the Bank off Greenock and all hands were paid off.

So far as any man who voyaged on deep waters could call a place on land his own, Frank Harvey had come home.

Although he had been born and lived his childhood in Alloa, from the moment the mountains of Cowal behind Bute had opened clear of Arran as the ship ran up the Firth of Clyde he was filled with an excited consciousness that his roots lay here. Intuition became certainty as, having passed between Toward Point and Skelmorlie Bank into the upper Firth, he began to recognise the detail of places and the lofty snow-capped peaks that he had seen when, between leaving *Prince George* and becoming Mate of *William Mitchell*, he first visited his Grandmother still living in Port Glasgow two winters before.

He remembered some of his many connections in Port Glasgow and Greenock, but most of all he remembered meeting the beautiful and commanding Margaret Watson, sister of the ships' chandler who had provisioned *Prince George*.

He resolved to take his discharge and seek out his relations and a new life in Greenock.

PART TWO

Sail on Tyne and Wear

6

The Wearmouth Flemings

SHORTLY AFTER THE NORMAN CONQUEST, EDGAR ATHELING, the slain King Harold's Saxon heir to the English throne, waited in South Wearmouth with his sister Margaret for wind and tide to favour their flight from the Normans. There they first encountered the man who was to become her husband, Malcolm, King of Scotland, on one of his forays into England, and it was Margaret who gave her name to the Queen's Ferry across the Firth of Forth.

It was a nice historical parallel when some eight centuries later a roving Johnston with a Ferry Hill lineage took as his bride a Fleming girl from Sunderland, as the port of South Wearmouth came to be known.

The importance of Sunderland was originally founded on coal and lime, and it was principally with the latter trade that the early Scots community was concerned. It was shipped from the Wear in small vessels, mainly sloops of between 40 and 100 tons burden, some belonging to the quarry owners there and others to the east coast ports in Scotland.

In his day old George Johnston of Kirkcaldy was familiar with this trade, and half a century later John Harvey of Alloa too. It was to this community that the elder Robert Gibson first came to seek his fortune in England when he left Dunfermline in the 1720s.

Who or what the first William Fleming was we do not know. He was doubtless descended from one of the Scottish families which settled in the port of South Wearmouth when it started to grow at the beginning of the 17th century. In 1727 he had a son William who, as a boy living at Wearmouth Pans, grew up among the watermen, tending the small boats they used on the river for ferries, passage work, the salmon fishery and all the multifarious purposes of the waterway that was at the heart of the community. In his twenties young William Fleming was making enough of a living as a foy-boatman to be able to marry Cicily Jordan and raise a family. They were both members of the Scotch Presbyterian Chapel in Robinson's Lane which in 1727 had seceded from the original Scotch community's Chapel in the Cornmarket.

The foy-boat was a lightly built rowing boat for river work, widely used by the watermen of the Tyne and the Wear for many utilitarian services to shipping such as assisting in the mooring of sailing vessels, transferring shipmasters between ship and shore, and carrying supplies. About 15 feet overall, 4 feet 6 inches in beam, with a narrow transom stern, it carried a dipping lugsail which played its part in the furious races to be first alongside to sell the waterman's services to a vessel coming in over the bar, but it was reputed to be difficult to handle under sail. In order to lick the competition in hard times, some of the most enterprising of the watermen—Will was one of them—sailed well out to sea to pick up their ships , and let themselves be towed in.

Will and Cicily Fleming had six children between 1750 and 1764—William, Margaret, John, Hedworth, Anne and George. About this time the family crossed the river to Monkwearmouth, then beginning to expand along the shore opposite the old Pans as a result of the increasing commerce of the port of Sunderland. Its growth was later accelerated by the expansion of the English shipbuilding industry that followed the American War of Independence. The Flemings became well established there. Will acquired a coble, a somewhat larger boat than the foy, more suited to open sea work and, indeed, the primary craft of the pilots and inshore fishermen.

On the Tyne and the Wear the traditional northeast coast coble, varying between 20 and 25 feet overall, was built with a narrow transom and painted in red, white and blue. It was clinker built with sawn timbers, almost universally larch, built by eye without moulds. No boat is more distinctive. Its high, sharp bow and long flat floor were evolved to fit it for rough water and for beaching stern-first through surf. A flat-floored coble was stable and was considered best for strong winds; a rising floor gave speed and easy lines; and a round bottom was thought best in light winds. It was rowed with heavy, characteristic square-loomed oars; but chiefly it was sailed with a tall dipping lugsail and jib. Under sail it needed half a ton of shingle ballast which was stowed either in the bilge or in canvas bags.

There being no boom, the coble was sailed by the sheet because it was easier to spill the wind out of the sail in a squall than to luff the boat up. In fair weather the jib was set on a bowsprit which served in heavy weather to carry the double-reefed lug. There were at least two steps for the mast, which had a lot of play fore and aft when the clamp was on. By means of a wedge before and behind it the mast could be raked to the exact angle required, as much as ten degrees in a heavy sea.

Sailing to windward in a breeze the coble would carry more sail than she could run with, and it needed nerve and skill working the sheet so as to avoid shipping water over the lee gunwale while keeping her going her best in a blow. Owing to the tumble-home the water was thrown off at the shoulder and amidships, so that the gunwale often seemed to be below the waterline without shipping any water.

When sailing fast, the water-pressure made the rudder vibrate so much that the whole craft and mast quivered, and when sailing against a very heavy sea in a breeze of wind the boat travelled so fast that, for fear of splitting a plank, the sheet was eased as the lee bow fell upon the sea.

Easily beached and light to row, the coble would work to windward under canvas as long as it could show sail; but the dipping lug was a dangerous sail and the square-sterned coble was apt to broach to when running before a heavy sea.

Sometimes the long rudder could break up or be unshipped by a sea. The sail was generally lowered and usually dipped to tack, though in some localities it was not dipped in moderate weather. Cobles were always towed stern-first with rudder unshipped. For fear of a capsize the tow-rope was never made fast but held by the boatman.

The latter part of Will Fleming's life (he died at Monkwearmouth

Shore in 1780) was spent as boatman in the pilot service. He ranged far out to sea in his coble, sometimes as far away as Yarmouth Roads, to find ships. There was no shelter for the four or five men on board but a canvas tilt that could be spread over the mast when it was lowered, nor anywhere to stow provisions for the long periods they were at sea. For food, he and his companions had to depend on what they could get from ships they met. They cooked it in a firepot kept in place on top of the ballast stones.

Will's coble was the family's livelihood and from an early age his sons played their parts in working it and looking after it, each showing his particular aptitude and refining the elementary skills that were to be the basis of his livelihood in later life. His eldest, young William, was a natural seaman. He had an unusually keen instinct for wind and water. No lad of his generation on the Shore had more sensitive hands for mainsheet and tiller to make a coble fly, or was more daring to handle it to the limits of its capability.

In later life he found his livelihood in the speedy coastwise smacks out of Berwick where he eventually settled and had a son, another William, who became a Master in the express sailing packet service between Berwick and London in the early days of the struggle against the growing ascendancy of steam.

Old Will's second son John, born on the Sunderland side of the Wear in 1756, was a big, strapping lad whose fine physique was not matched by any great skill or intelligence. Of all the boys, John was the least interested in boats and the sea, but he was just the right age—as well as either stupid or unlucky enough—to be caught by the Press and shipped off to serve in the Navy on the Western Atlantic during the campaigns of the American War of Independence.

Forcible impressment was the normal way of augmenting the manning levels of the Navy from the days of Henry II until Victoria's accession. It was totally abhorred by the classes of society most grievously affected by it, and wholly accepted by the upper classes despite its dubious legality. Ashore it was carried out by Naval Press Gangs made up of strong, wild thugs, not always naval seamen but invariably under the command of a commissioned Naval Officer with a couple of Midshipmen. Usually they were armed with oak cudgels and cutlasses. Sometimes they also carried pistols and blunderbusses when the going was tough.

Afloat the Navy had an extensive organisation to prevent merchant seamen getting ashore from homeward bound vessels (the impressment of outward bound merchant seamen being usually prohibited) and to transfer them to men o' war waiting in immediate need of manning.

Sloops and frigates patrolled offshore, the former searching continuously between Beachy Head and The Downs where shipping lanes were concentrated; while closer inshore and in the big estuaries the Press Gangs hunted in tenders, and sections of these gangs patrolled the beaches and creeks in small boats.

Although officially some categories of merchant seamen were exempt, none, whether ashore at home or abroad, or even afloat in their own ships, were ever free of impressment into the Navy in wartime.

Masters, Mates, Boatswains and Carpenters of merchantmen over 50 tons were exempt while they were afloat or on ship's business ashore and the whaling trade enjoyed special exemption. Impressment from east coast colliers was always subject to restriction, sometimes being completely prohibited, and there were exemptions for seagoing apprentices and for all seamen in the first two years of their seagoing career. But the Press Gangs were often a law unto themselves and certificates of exemption counted for little when the gangs were falling short of the quotas they had to fill.

During 1776 and 1777 the coal trade was depressed, unemployment was high and wages were low. Whenever there was unrest the coal owners were not above using the Press as a means of disciplining the seamen of the collier fleet, and that winter the Press was particularly active at both Shields and Sunderland with a ruthlessness and on a scale that were not unconnected with this unrest. And not only seamen, but watermen and even landsmen were taken when the Press was hot.

John Fleming was one of the unlucky ones. Forcibly seized, he was dragged away and flung into an Admiralty Tender where, with a hundred others, he was confined in the hold under gratings with a Marine Guard at the hatchways, muskets loaded and bayonets fixed. In conditions akin to those of a slave ship he spent five nauseating days in passage to the Thames Estuary. Off the Nore he was transferred to a receiving ship to await drafting to a man o' war. Now he began to feel the real horrors of naval discipline.

Life afloat, be it in the merchant fleet or the Navy, was naturally harsh; but what made a hell of life afloat in the Navy was the absolute loss of liberty; of freedom to speak or to act; the habitual exercise of physical violence; the hideous overcrowding below decks accompanied by universal drunkenness and lechery; and the commonplace rituals of flogging of a degree of savagery that it is difficult to even imagine today. Brutality was deemed to be essential for discipline, and the men were controlled by the rattan, the rope's end and the cat under the shadow of the Marines' muskets. Britain was a maritime nation but few ever deliberately chose to serve in her Navy.

Although he spent much of his time on the eastern seaboard of the American colonies or in the West Indies, John Fleming did not take part in any major engagements of the war. He served in one of the frigates sailing endlessly backwards and forwards across the Western ocean or up and down the eastern seaboard, escorting convoys in support of the land operations—to New York, St John's Newfoundland, Quebec, the Delaware River, Charlestown, the Leeward Islands, Jamaica—a constant, monotonous round of seagoing duty, year in and year out, with only the occasional excitement of an encounter with an American privateer.

It was a wearisome enough life for any man, and when his ship paid off at Portsmouth on the return of peace early in 1783 he swore a great oath never to go to sea again.

Back home at Monkwearmouth he found that there was not much scope on the beach for an old Tar at a time when the Navy was discharging men by the thousand. His father was dead, his elder brother William was away in a Berwick Packet and his younger brother George, the liveliest of the lot of them, had by a natural dominance of character assumed the role of head of the family.

In the year before John was pressed, his elder sister Margaret had married a minor merchant, Robert Topliffe, who had done well out of the war as a small shipowner, so that the jobless John found them a mite too superior for his taste. His brother, Hedworth, was concerned with the prospect of his impending marriage to Ann Robison, and their younger sister Ann was preoccupied with the impending birth of her first child by her new husband William Lawson, a mariner. So, although his family welcomed him warmly enough, John felt isolated from them and kept himself very much to himself.

Gradually he found casual labouring jobs for which his sturdy physique made him well fitted. He worked for a while on the demolition of the old south pier, but it was the construction of the new North Pier at Monkwearmouth that was soon to provide him with a steady livelihood as a labourer for many years. He married Jane Tate who bore him 14 children between 1790 and 1830.

From their earliest days the younger Fleming boys, Hedworth and George, showed a talent for working in wood and iron. They were never happier than when making repairs to their father's coble. Of the two, Hedworth was the more dedicated to sheer craftsmanship, while George shared something of his eldest brother's talent for seamanship.

They were fortunate to be born in an era when the northeast was just beginning to establish its position at the top of the ship-building league. Until they were challenged by the British North American colonies in the 18th century, the Dutch had been the supreme shipbuilders; but now, following the creation of the independent United States of America, the northeastern ports of England moved in to fill the gap. Sunderland's rapid rise in importance and the growth of Monkwearmouth Shore date from this time.

Both Hedworth and George served their apprenticeships in one of the small ship-building yards on the Shore. Having learned their trade as shipwrights, however, both boys found it more congenial to serve afloat as ship's carpenters.

In a sailing vessel the carpenter occupied a special position of trust and responsibility. He was a man of wide skills, being also a blacksmith, a block and spar maker, a joiner, a caulker, a boat-builder and, incidentally, a sailor. Together with the Mate he superintended the task of careening ship.

A vessel of even 300 tons burden could be careened without complicated machinery or elaborate dockyard facilities, and it was the carpenter's responsibility to give the vessel sufficient internal strengthening as well as to lash spars to act as levers between the vessel and the quay. Purchases attached to the masthead were anchored to the quay; all hands heaved and, as the purchases were shortened, the hull, pressed up against the spars, canted over until the keel was clear of the water.

There was also the regular process of 'breaming' which fell under the carpenter's eye. The vessel was allowed to dry out on a hard shore while the crew cut fuel for fires; then brands of burning reeds were used to remove marine growths on the hull and cauldrons were heated to re - tar as needed. These activities, and shaping spars from baulks of timber; searching out leaks and curing them; and spiking, strapping, and generally cosseting his vessel were all jobs for the carpenter. The pumps, too, were under his charge, and the slings and parrels and, when they were superseded, the iron trusses and goosenecks upon which the yards were balanced, and the working of the iron blocks and sheaves and the maintenance of masts and spars.

Unlike anyone else in the ship below the rank of Mate, the Carpenter found his own work. He was not called out with the watches to handle canvas and rope, but the best—and George was such—prided themselves as being as able and alert aloft in a blow as any of the hands. His 'shop' was a place into which no-one but the Master, and possibly the Mate if he were of sufficient stature, had right of entry.

George and his wife were a lively couple who did much to cement the Fleming family that gradually dominated the congregation of the Shore's Scotch Church. By the turn of the century the Flemings had become a sizeable tribe closely interwoven in the maritime community. The records of their growth are spread among the Presbyterian Chapels, two in Sunderland and two in Monkwearmouth.

The earliest recorded were the six children of William and Cicily Fleming baptised in Robinson's Lane chapel from 1750 onwards. The first two children of Margaret Fleming and Robert Topliffe were baptised at Spring Gardens Chapel in Sunderland, but the next four were baptised in the Scotch Church at Monkwearmouth as were the children of Margaret's brothers and sisters and many of their descendants. Margaret was the real matriarch of them all.

When she was 11 years old she had taken on the duties of her mother Cicily who died shortly after George's birth. She and her husband Robert Topliffe did much to provide opportunities for their many nephews to go to sea and learn their trade as ship's boys. She attracted the deepest love and affection of all her younger brothers and sisters, each of whom gave her name to her eldest daughter.

Her brother John went his own way in religion as in all else; there was an independent Chapel in Dundas Street on Monkwearmouth Shore, the register of which contains a list of the names of all 13 of the children of John and Jane between 1790 and 1813.

In 1806 John Fleming celebrated his fiftieth year with the birth to Jane of his eleventh child, Joseph. Most of the tribe was present at the baptism with, of course, Joseph's nine living brothers and sisters: William, Margaret, Mary (13 years old and in a few years to marry a Tyne mariner and settle at North Shields), Jane, John (a future shipmaster), Ann with her mariner husband William Lawson, Elizabeth, Hannah and Daniel (another future shipmaster).

George and Jane Fleming were accompanied by their brood: John (to be a shipmaster), Margaret, George and Jane (who like her parents was a person of unusual warmth). And of course the Topliffes were there with their two youngest, William and Robert Junior. Hedworth and Ann Fleming were accompanied by 13-year-old Margaret and William, their youngest.

Hedworth's Margaret did not find a man to marry until she was nearly 30, and then it was to marry beneath her family's station, for she married James Campbell, a keelman. These were a colourful, fiercely independent, aggressive band of men who played an almost legendary part in the history of the Tyne and the Wear for nearly five hundred years until their extinction in the 19th century.

A hard drinking, gambling, turbulent lot, the keelmen were the subject of numberless ballads. The 'Keel Row', the most widely known of them, was a popular melody on Tyneside before 1760; but the keelman is most graphically described in two verses of 'My Bonny Keel Laddie' (keel, derived from the Anglo-Saxon ceol, a ship, is pronounced as two syllables):

> 'My bonny keel laddie, my canny keel laddie
> My bonny keel laddie for me, oh!
> He sits in his keel as black as the de'il
> And he brings the white money to me oh!
>
> The canny keel laddie, the bonny keel laddie,
> The canny keel laddie for me, oh!
> He sits in his huddock and claws his bare buttock
> And brings me the white money, oh!'

A clannish community, their calling descended from father to son and they usually married only with their own class or the pitmen. They were bound by a yearly bond to their employers, the hostmen or coal merchants who owned the keels. These were oval, flat-bottomed barges built to carry the coal from the dykes or staithes in the upper, shallow reaches of the two rivers to the colliers in their berths in the harbour, and sometimes even to colliers at anchor outside the bar.

The keel kept its almost mediaeval form for more than four centuries: the length was about 40 feet, the beam 19 feet amidships to give stability; the laden draft less than five feet (shallower on Wear than on Tyne). It was massively built. The floor of the hold, supported by huge beams, was barely two feet below the gunwale, so that the coal, piled high above the deck level, had to be kept in place by boards. Although the keel set a single square sail and could be propelled by a pair of oars and steered by a twenty-five foot swape over the stern, its main motive force was the tidal stream. The size of keels was standardised in the 17th century at eight Newcastle chaldrons, just over 21 tons. The trade sold coal normally by the keel-load and described the capacity of the colliers by the number of keels they could load on board.

Like most keelmen on the Wear, James Campbell managed his craft with only a boy (invariably known as the 'PD') to help him. A team of coal casters was employed to shift the coal from his keel to the collier. On the Tyne, on the other hand, the crew of the keel, comprising three men and the PD, cast the coal from the keel onto the waiting vessel themselves.

In the early 1820s the Sunderland keelmen readily accepted a new practice of loading coal in iron tubs of one chaldron capacity, which were then lifted by crane into the holds of the colliers where labourers emptied them. The Tyne keelmen, on the other hand, did everything possible to resist the onslaught of such new technologies and practices.

7

The London Coal Trade

THE INDUSTRIAL REVOLUTION OF THE 19TH CENTURY rested on two pillars, iron and the coal trade of the northeast. The latter was dominated from time immemorial by Newcastle and its Company of Hostmen, the Guild of Coal Merchants created by Royal Charter in 1536 which subsequently acquired a virtual monopoly of the government of the Borough and a leading role in the Commission created to exercise jurisdiction over the Tyne navigation.

Although sporadic domestic coalmining began on lower Tyneside and lower Wearside in the 13th century, it was another 300 years before markets were developed in London and other towns round the basin of the German ocean; nevertheless until well into the 18th century the great bulk of shipping engaged in the coal trade belonged not to Newcastle but to Whitby, Yarmouth, London and King's Lynn.

The ultimate predominance of the coal trade between London and Newcastle came about mainly because both ports were on navigable rivers serving large hinterlands in an age when there was no means of transporting bulk commodities overland. London looked to the northeast and to Newcastle in particular (the growth of Sunderland's importance came later) not because there were no other sources of supply, but because at Newcastle there were abundant outcrops close to navigable water and to facilities for shipping.

In the early centuries of the trade, coal was carried by miscellaneous ordinary trading vessels, but later, ships were built primarily for the trade and a fairly standard pattern emerged: first the cat or hag boat, then the brig and the snow. These were served by the keels, which brought the coal down to the waiting ships, having been

A Snow at anchor

loaded at the coaling staithes on the river bank a short waggonway from the pit-heads.

After about 1720 the waggonways were driven further and further inland, especially in the middle Wear Valley and the northwest Durham plateau where deeper seams came to be mined as the technology of steam-pumping to keep water out of the lower levels was developed.

In the 1770s the handsome, spacious ship-rigged cat boats with their great big stern windows were being superseded by the flush-decked, square-sterned little brigs, often with no bulwarks or even rails to protect the deck amidships. Very soon the brig proved itself to be the handiest of all square rigs and came to dominate the coasting trade. She also showed herself to be the most powerful and efficient rig for deep sea work., for she carried as much square sail as a barque half her size again, and needed fewer hands.

It was possible to sail a brig with six hands, but eight all told was more comfortable. The old Whitby ship-rigged cat boats needed at least ten, plus a couple of apprentices for good measure. What the brig lacked in speed she made up in carrying capacity. Her seaworthiness and manoeuvrability even when deeply laden were demonstrated by the fleet's ability to thrash through the open seas between the northeast ports and the London river in all weathers, and deliver the cargoes of coal through one of the most crowded waterways in the world deep into the heart of the capital.

By the early 1790s when George Fleming was appointed Master of *Jenny*, one of his brother-in-law Topliffe's vessels in the London coal trade, the seasonal character of the trade had been somewhat modified and winter sailings had risen to about one third of the summer peak. *Jenny* was a typical Geordie collier brig, not to be confused with the larger, more elegant and finer-lined Baltic Traders and South Spainers which, although they carried coals far and wide, engaged only spasmodically in the shuttle between Thames and Tyne or Wear. She was apple-bowed, straight-stemmed with not even a pretence of fiddle-head, with a flat transom stern. 'Built by the mile and sawn of by the yard' she was heavily rigged, crossing a single topsail and carrying a full set of stu'nsails.

Jenny and her like operated in large fleets and passage-making was always a race to be first to market. On the first leg of the passage southward from the Wear there was usually a fetch with an offshore wind as far as the lee of Winterton Ness or Yarmouth Roads. In a northeaster embayment in the Wash was an ever present hazard which, in a gale, meant almost certain destruction. Robert Topliffe, as owner of

Jenny, issued orders to his Masters never to sail in December, January or February with the wind southward of southwest of eastward of northeast.

In March and October George could use his own judgement in consultation with a local pilot. For the rest of the year he was under an obligation to sail with the first tide whatever the wind direction so long as *Jenny* could carry whole canvas. But it was the middle passage between the Humber and Yarmouth that could turn into a very dangerous region when the wind blowing hard onshore made conditions too bad for anchors to hold. The shallow Wash, with no shelter to leeward and escape to windward impossible, could be a death trap.

From Yarmouth it was a long slog in the prevailing wind, often a beat the whole way to the Pool of London. At every turn of the tide against him, perhaps first in Yarmouth Roads, next in Hollesley Bay and after that the Swin, he dropped the huge wooden-stocked anchor. Smart anchor work was one of the hallmarks of the collier, and the windlass had to be simple and reliable; but the labour that it demanded for swift anchor work in streaming tides was back-breaking.

At night, even in the relative calmness of the prevailing offshore winds, navigation along the east coast was not without hazard. Orfordness, first lit in 1603, Lowestoft in 1608, Winterton in 1615 and Cromer in 1764, were all still fired by unreliable coal baskets while George Fleming was serving afloat. The first oil lamps were not installed until 1792 at Cromer and Orford, followed by Lowestoft in 1796. And they were privately operated.

Cromer, for example, was established and paid for by the coalmasters of Ipswich, Yarmouth and Newcastle. The very first lightship was placed on the Nore Sand in 1732. It was followed four years later by the Dudgeon, and by Happisburgh in 1789. At Tynemouth the Castle Light Tower, rebuilt in 1744, was fired by coal until 1802, when candles were used in a revolving copper lantern.

The Thames that George Fleming could feel his way into by instinct and leadline was a sprawling, cluttered waterway with critical shallows at and above Barking Creek that restricted the passage to drafts exceeding nine feet. Barking then had much of the character of a coastal port. The next shelf was just above Woolwich where George picked up his river pilot and towed his skiff astern. *Jenny* would have to anchor in the deeps above the Woolwich Naval Dockyard if she could not make one of the higher holes off Blackwall or Deptford on that tide.

The depths of the upper reaches of the Thames broadly matched the depths on the bars of the Tyne and the Wear, setting a limit of 14 or 15 feet on the draft of colliers which governed and eventually standardised their tonnage irrespective of their rig. A little over 200 tons burden became the norm around which the whole process of loading, discharging, lightering and storage became formalised.

The skippers of these brigs navigated with consummate skill the twists of the upper reaches of the London River where almost any wind was bound to be foul somewhere (the loop around the Isle of Dogs, for example, involves a change of course of 180 degrees). As the channel was too narrow for tacking they had to resort to 'backing and filling'.

The brig was sailed as closely on the wind as possible. When the river bank or shoaling water blocked further advance, the yards were braced round until the sails were aback, whereupon the brig stopped and gathered sternway, still lying across the current. The sternboard was continued across the river, the current all the while carrying her on her way. Some of the Geordie brigs were said to sail better stern-first. Near the opposite bank the yards were braced round again to move her forward on the original tack. The pattern was continued until a new bend of the river enabled her to reach clear.

In company with a score or more of other colliers amongst the hazards of drifting lighters, ferries and fishing boats, the operation needed a steady nerve, superb timing and a handy crew.

Operating at such close quarters without benefit of established rules, and at night without benefit of lights, even superb seamanship could not entirely avoid serious collisions. In later years the Bentinck boom was brought in as a standard feature of the collier rig to simplify sail handling while manoeuvring in narrow waters. This boom was bent permanently to the clew of the forecourse; it had a single sheet at the centre of the boom which ran through a tackle to a ring-bolt in the deck just forward of the mast and could pivot round the ring-bolt as the yard was braced around. It did away with the need for handling tacks and sheets every time the vessel went about and made it possible to retain the service of the forecourse which otherwise would have to be furled or clewed up.

Jenny's final anchorage was the Pool of London. Permanent moorings for colliers were not to be laid for another forty years, hence the importance to the Masters of being in the van of the fleet to get the best berths for anchoring in the foul mud of the Pool. Here George supervised the stowing of sail with full bunt and clean yardarms before being rowed ashore in the waterman-pilot's skiff that had been towed up from Woolwich.

His first job ashore was to deliver at Custom house the Cocket from the Customs Officer at Sunderland where he had loaded the coals. He and his coal broker, known as the 'crimp', arranged for the payment of the King's duties and then went off to register the Fitter's (factor's) Certificate at the City Cocket Office where the City's dues were paid.

When the Customs Officer and the Lord Mayor's Officer had given George a warrant for the Coal meters authorising them to deliver the cargo, the business of selling coal could begin.

The coal market, formerly at the head of Billingsgate Dock on a ground called Room Land, was by now housed in a Coal Exchange on the north side of Thames Street opposite Smart's Quay. In the old days of the Room Land market the Master himself would have negotiated the sale, but now George left it to the 'crimp', who was variously a broker, undertaker or wholesale dealer, to find a buyer on the Exchange and to arrange for the meter, labourers and lighters so that unloading could begin.

When the cargo had been discharged the meter endorsed the customs warrant to show the amount of coal delivered as the basis for the crimp to pay the meter and the customs and coal duties. The factor also paid miscellaneous bills on behalf of the Master, advanced money for the uses of the ship such as provisions and stores, and paid the undertaker for labour. Then the Fitter finally settled with the shipmaster.

There was no official regulation of shipping on the Thames before

1799 (but, surprisingly, a system of 'turns' existed formally on the Tyne from 1766 onwards). For every Master it was a case of Devil take the hindmost among perhaps 200 ships in the Pool and scores of lighters milling around them. Coal was delivered to the lighters by a gang of nine 'whippers' provided by undertakers who were usually publicans.

Between Tower Hill and Limehouse there were at least seventy taverns along the waterfront where men who wanted work gathered every day and, of course, priority went to those men who spent most of their wages in the undertaker's house. The whippers were a savage, dissipated, lawless class of men whom the publicans exploited mercilessly. Of the whipping team, four men in the hold shovelled the coal into baskets; four more men above deck whipped the baskets out of the hold making use of a ladder-like 'way'—usually four old oars set to form steps upward of about a foot. As they mounted the way they hauled on the whip tackle that raised the basket, then lowered the basket by jumping down onto the deck. The foreman standing on a plank across the top of the hold guided the basket so as to shoot its contents into the weighting box, from which the coal was funnelled into the lighters lying alongside.

The normal rate of delivery was two score chaldron a day. The coal-whippers were capable of delivering three times as much but the buyer was only obliged to produce lighterage for twenty-one chaldrons a day. The shipmaster on the other hand was subject to considerable financial penalties if unloading was delayed. There was therefore a lot of scope for offering and accepting presents and the exaction of 'despatch money' to accomplish the quick turnaround that every Master desired.

When unloading was finished George's one desire was to catch the beginning of the ebb for the dash back to the north, but his light, crank ship needed first to be ballasted with gravel. Once again it was to the publicans that he had to turn for a contract for his ballast. The cleanest was lifted from the bed of the river below Woolwich by teams of half a dozen ballast-getters and was carried in lighters to the Pool where other teams of ballast heavers shovelled it into the ship's hold. A really cranky ship needed as much as sixty tons to stiffen her; a stable brig like *Jenny* needed much less. Publicans contracting for both coal-whipping and ballast did very well financially in George's era, many becoming shipowners themselves. Several had family ties in the north-east ports.

George Fleming's command of *Jenny* did not last long. The huge expansion of ship-building in the northeast set off by the maritime phase of the Napoleonic War stimulated a great demand for shipwrights in Monkwearmouth, and his growing family helped pull him away from life at sea. He was a fine seaman among a breed of the finest seamen in the world; but to be sole Master under God in the face of hostile elements required uncommon qualities of confidence and endurance which had stretched him to his limits. If truth be told, he found life less wearing in the honourable trade of shipwright.

His brother, young Joseph Fleming, like his kin and cousins before him was bred to the sea. When he was ten he officially became one of the lowest of all maritime creatures, a ship's boy. In the ship's papers this miserable but indispensable brat was, like the Master, shown in his own category separately from the crew. But if the Master lived in the hell of lonely responsibility the boy's life was a hell at the opposite end of the scale.

The youngster afloat was treated to the point of cruelty, totally at the mercy of the Master, even the best of whom believed in the efficacy of harsh treatment as the only proper training for the toughest of callings. And he was the butt of all the ill temper of the ship's crew.

His duties were many and varied. He minded the ship while the crew was ashore; he sculled the Master about in the Jolly-boat whenever needed, a job calling for no mean skill in the swirling, wind-whipped Tyne or Thames at dead of moonless night. But Joseph had inherited all the right skills. After three or four summer voyages he returned to the nautical school in the winter, and the following year began a seven-year apprenticeship in a Baltic trader.

As ship's boy he had learned as much of the ordinary seaman's trade as his strength allowed: drawing and knotting yarns, making spunyarns, foxes and sennet, loosing and furling the light sails and reefing them, making all the common hitches, bends and knots, rigging studdingsail booms and reeving all their gear, and even taking the wheel in light weather.

As apprentice his status was initially no higher than that of boy, but as his strength grew so the range of his duties broadened until there was no sailor's task he could not perform.

During this period of his life Joseph Fleming served under a young Mate from North Shields, George Hutchinson, who came ashore one Sunday to worship at the Scotch Church at Monkwearmouth and was pointed out by Joseph to his family, including his sister Mary, then in her early twenties. One thing led to another and in due course Mary married Hutchinson and set up home in North Shields. Whenever his ship lay in the Tyne Joseph stayed at his sister's and worshipped with her at the Scotch Church there. Its congregation included William Johnston from North Queensferry who had settled at North Shields with his wife Helen Nelson, to beget a numerous family while plying in the Baltic Trade.

George Fleming and William Johnston were accustomed to manoeuvring under sail to fetch up at anchor in a crowded pool, but when Joseph Fleming was born there were 4,500 vessels in London's coal trade. The Corporation was forced to initiate a series of bylaws under the 1799 Port Act, which created its first harbourmasters and initiated the great system of docks. It ordered that colliers must lie in specified moorings in tiers of not more than fifteen ships each. Nevertheless there was no positive control of flow until 1821, when a flag hoist was introduced to stop colliers at Blackwall and Greenwich until there was room in the Pool.

By the time Joseph was a young Mate the fleet had increased to 7,000 vessels, almost 2 million tons, and the coal trade was insisting that it could not meet its obligations unless the capacity of the moorings in the Pool was expanded to cope with 250 sail. Although vessels were being held at Blackwall and Greenwich until the Pool had room for them, the only effective flow control was the tide. So long as sail was the motive power, navigation through the Pool was a matter of arrival on the flood and departure on the ebb.

Steam propulsion changed things dramatically. The largest vessels, the steamers, came and went independently of the tide, and owners aware of the need for high rates of utilisation in order to get a return on their costly new technology demanded vigorous management and superintendance of the Pool and its approaches by the Port authority.

By the time old William Johnston was sailing *Mary* into the London River in 1835 everyone was complaining about the uselessness of the harbourmasters, the state of traffic in the river, and the new Dock companies' infringement of mooring space traditionally allotted to the collier fleet. The likes of William and Joseph were complaining about the damage they received from steamboats or vessels towed by steamboats; unnecessary detention at Deptford, Greenwich and Blackwall; being harassed by orders to move off the Red House and the Dockyard at Deptford or off Greenwich because steamboats wanted to go to and from the piers all day; and being incommoded by the Dock company's regulations at the West India Dock, the Commercial Dock or the Greenland Dock. At the Folly House they were cursed by the Masters of the steamboats, at Blackwall and at the second entrance of the West India Dock and at the East India Dock they were harassed by the Dock Masters, and if they lay in an intermediate space they were in the way of the Brunswick steam wharf. They felt that they had more damned regulations as well as more expenses to cope with than any other trade in the Port of London.

In about 1834 the Coal Factors' Society established an office in Gravesend, where they appointed Mr James Read as Resident Clerk for receiving and forwarding to the Coal Exchange all papers sent to him by Masters of the colliers. He entered and numbered them in order of arrival. The Factors sent a Circular to Masters recommending them 'immediately they come within the limits of the Port of Gravesend (whether by day or night), and not before, [to] send their papers to Mr Read at the said office, which is situated in the lower part of the town, close to the river, affording easy access to ships' boats at all times, and by a punctual observance of this regulation all ships will secure a certain and regular turn on the list for a Meter. It is strongly recommended that Masters of ships do not give their papers to any parties pretending to forward them to London as this practice has caused much irregularity and detention.'

A contract was placed with a steam packet company for transporting the papers from Mr Read's office to London. There was nevertheless still much to aggravate the shipmaster between Gravesend and the moment of unloading at the Pool, for at peak times there might be as many as 700 sail in that stretch of river and there were only eight named tiers of moorings for the exclusive use of colliers in the Pool, five on the north bank and three on the south, providing for no more than 147 vessels. Rarely were there fewer than 100 colliers waiting between Limehouse and Bugsby's Hole for room in the Pool, and sometimes more than 200. In the winter of 1838/9 Johnston's *Mary* was detained for twenty-three days between Gravesend and Greenwich awaiting orders for the Pool.

Because Mr Read moved his paper more swiftly than the Harbourmasters could move the traffic, Meters were appointed and kept waiting for ships that were prevented from coming up to the Pool. There was moreover no system for advising Masters of their chances at Market, and until or unless the coal was sold the vessel could not be unloaded.

By the time Joseph Fleming had become Master of *Elizabeth Adnett* the regulation of traffic on the London River had become quite sophisticated. From about 1842 the system of control involved three subordinate Harbourmasters working closely with the Coal-meter's office. One looked after the Pool itself, where eight tiers between London Bridge and Limehouse now accommodated 243 vessels. The second controlled the reaches from Limehouse to Barking. The third handled the stretch from Barking to Gravesend and below. Lying upstream from Northfleet as far as Blackwall there were seven numbered sections of tiers at which colliers lay awaiting their turn for the Pool, having been allotted their section by the Harbourmaster at Gravesend. There were special arrangements for giving priority of berths in the Pool to the growing number of vessels whose cargo had been sold prior to reaching the London River.

During the decade between 1834 and 1844 the London Factors

conspired with the northern coal owners to limit the vend in order to improve their profits. The limitation of the vend was effectively terminated at the same time as the breaking of the coal strike of 1844 and the abandonment of the rigid application of the turn system for moving colliers on the London River. The ancient practice of selling the coal to the shipowner in the north was on the way out as he became increasingly merely a carrier, and the period in the mid forties when Joseph Fleming was sailing *Elizabeth Adnett* into the Pool was one of great change in the London coal market.

Before the depression of the early forties forced Captain William Johnston to downgrade to Mate of the broken-down collier *New Eagle* and then into retirement, virtually all of London's coal came into the Thames by sail from the north. In the early fifties, by the time his son James was a young Master of the smart Baltic trader *Pomona* only secondarily engaged in the London trade, coal was being brought in by rail from inland areas and from the west, and steam screw colliers, aided by the development of mechanical methods for handling their cargoes, were beginning to dominate the carriage of sea coals. A sailing collier might make perhaps ten voyages to the London River in a year, each with a couple of hundred tons of coal; but the steamers, even of the first generation such as the *John Bowes*, could make upwards of 30 voyages each carrying between 400 and 700 tons, and needed only a single day for discharge.

Moreover, less coal was being consigned to the London Coal Exchange. Wholesalers were buying more of it directly from the coal owners in the north, with the result that steamers not only won priority for unloading facilities but also, because of the growing demand by various new large-scale undertakings such as gasworks, were able to land their cargoes at points where delays could be avoided.

Steam propulsion changed things dramatically and the owners of the steam colliers were vociferous in their demands for vigorous management of the Pool and its approaches. When John Fleming sailed his first command, the collier *Elizabeth*, into the London River in the winter of 1860-61 the old turn system had practically disappeared in the Pool.

8

The Tyne Brigs

NORTH SHIELDS WAS A THRIVING FISHING HARBOUR even before the reign of that 'Hammer of the Scots', King Edward I, deriving its name from the Scots 'shieling', a hut. Its harbour comprised a one-and-a half mile stretch of the Tyne, from the North Shields Low Light almost a mile within the bar, to the lower end of Jarrow Slake. It was clear of the swell of the open sea, but the shoals and the varying curvature and width of the river caused considerable variations in the tidal stream. Although the stream at half ebb was nearly 4 knots in mid tideway opposite the quay at North Shields, easterly winds which blew straight into the harbour could greatly hinder vessels, particularly those of large draft: for by the time a deep laden vessel had dropped down on the ebb against the wind, there was as like as not insufficient depth of water to let her pass over the bar.

For hundreds of years the monopolistic control exercised by the Corporation of Newcastle and its Company of Hostmen over everything that happened on the navigable Tyne stunted the growth of North Shields, as it did that of South Shields across the river. Enactments at the end of the 17th century, however, enabled the two settlements that shared this superb natural harbour to develop independently.

North Shields, strung out along the river bank below the bluffs, grew thereafter on its sea fishery, which reached as far as Greenland and the Davis Strait, on its specially profitable trade in salt (in the making of which South Shields was at one time pre-eminent), some ship-building and ship-repair, and the general coastal and river traffic. It was the more clandestine aspects of the latter, and its prospects of growth that had attracted the Dunfermline Gibsons and George

Johnston of Kirkcaldy. The wealth from these activities enabled the planning and building of the New Town between 1750 and 1820 on the level ground above the high bank which had constricted the slum-like Old Town to the narrow strip on either side of Low Street. William and Helen Johnston, like George and Mary Hutchinson, were able to move out of the rat-infested wooden hovels there, where the whole trade of the harbour had been carried on, into the prospering community of younger shipmasters in the superior brick terrace houses of the Upper Town between Walker Place and Dockwray Square.

As long ago as the 15th century the conservation of the Tyne had been vested in Newcastle Corporation, abetted in this work by the society of Masters, Pilots and Seamen of the Trinity House of Newcastle-on-Tyne. For centuries the Corporation neglected its own river and fought hard to prevent the people of Sunderland from improving theirs, as well as the two Shields from exploiting their fine natural harbour. In the end their failure to improve the navigation of the Tyne contributed materially to the rise of Sunderland as a coal and shipbuilding port during the 18th century. Parliament eventually removed their dead hand in 1850 by creating the Tyne Improvement Commission.

Although at certain states of the tide the Tyne was navigable by vessels of as much as 250 tons burthen for the ten miles from its mouth to Newcastle, at some places upstream of Shields the depth at low water was barely four feet. The navigation of the river was therefore determined entirely by the rhythm of the tides, the flood lasting five hours and the ebb seven; it was high water at Newcastle bridge about an hour later than at Shields. The average rise of spring tides was just under 12 feet, slightly over 7 at neaps. In the early days of steam William Johnston and his contemporaries took pleasure in scoffing loudly at Masters of steamboats drawing a mere three feet of water lying aground in mid-channel on the last quarter of the ebb, waiting purple with choler for the flood to refloat them.

The second hazard of the river was its sinuousness between high banks, with great, sharp bends at Whitehill Point, Hebburn Quay and, most notorious of all, the dangerous Bill Point where a huge wall of land jutted out into the river like a beak. Ships of moderate size frequently ran aground in trying to pass this point, and the bank was so high that vessels rounding in opposite directions could not see each other before a real collision risk had developed.

But the third, the most fatal hazard, was the combination of the sand bar across its mouth with the vicious Black Midden rocks to the north and the shallow Herd Sand to the south, which made the inward passage of vessels extremely dangerous when the wind blowing from the eastward kicked up a heavy sea. A vessel coming in on an easterly blow and just losing her ingoing tide was not only landlocked but was also faced with a savage, steep wind-against-tide sea at the bar. A vessel crossing it might easily broach to if lightly laden, or if deeply laden might strike the bottom and damage her rudder. In either case, rendered temporarily unmanageable she was almost certain to be driven either by a northeast wind onto the Herd or by a southeast wind onto the Black Middens or the rocks at the foot of Tynemouth Castle.

The narrow safe line of approach over the bar, which at best had a depth of only six feet at low water springs, was first marked when Henry VIII licensed the erection of two stone towers at North Shields, the High and the Low, aligned with the channel and lit by candles. The stone towers were replaced by wooden ones in 1658, but 14 years later another stone Low Light was built within the Prior's Fort. The High Light was rebuilt in 1727 to provide better alignment with the shifting channel. By 1805 the channel had moved out of alignment again, so new towers were built in the following couple of years.

From year to year many lives were lost in shipwreck at the mouth of the Tyne, often in full view of the hushed, watching, helpless people of Shields and Tynemouth. In the winter of 1789, following the wreck of *Adventure* of Newcastle on the Herd Sand, attempts to construct a special purpose lifeboat were given serious encouragement for the first time. Thousands of people had watched as, one by one, the crew of *Adventure* had dropped from the rigging, exhausted by cold and fatigue as she lay stranded in the midst of tremendous breakers near the entrance to the harbour. As usual, no men could be prevailed upon to go to their assistance in any boat or coble of orthodox construction. Deeply affected by what they had seen, six influential South Shields men collected funds to finance the design and construction of a boat fit to survive the dangers of shallow broken water. Eventually they chose the design of Henry Greathead, the builder of the first Lifeboat which on its first operation saved several men from a wreck in the offing at South Shields on 30th January 1790.

North Shields became the home port of William Johnston of Inverkeithing following his marriage to Helen Nelson in 1816. There he sired his small dynasty of shipmasters that lasted a hundred years. He and Helen joined the High Meeting of Protestant Dissension, known generally as the Scottish Church, on Bankside at North Shields. In its registers are recorded the baptisms not only of their own children, but also of their grandchildren and great grandchildren.

Their firstborn Isabella was baptised in 1817; their first son James—the author's great grandfather—was born in 1819, and thereafter five daughters and two Williams (1831 and 1833).

In 1820 when he acquired his first command, the trading schooner *Elizabeth and Anne* out of Montrose, he was still a young, good looking, zestful and thrusting Scot keeping himself ahead in the competition for berths even if it did mean quitting, for the time being, the Baltic trade for the lime trade between Sunderland and the northeastern Scottish ports.

After serving for a year or two in this vessel he won for himself sufficient of a reputation as a hard, reliable Master to find continuing employment in the fleet of collier brigs working out of the Tyne. His last command, which he held for many years until 1842, was the 280 ton brig *Mary*, owned by James Gibson, scion of old Andrew Gibson the Dunfermline carrier.

William followed his calling with a degree of success which, in the years of social and industrial turmoil and depression which followed the end of the Napoleonic war, suggest that he possessed patience, physical toughness, professional ability and a proper degree of gregariousness amongst people who might be able to help keep him employed. That he spared himself little in the endless conflict with brute men and the elements is evidenced by his being completely burned out by the age of fifty, after which he took ten years to die of a broken mind in a remorselessly wasting body.

The brig was by now dominant in the short-sea trades. Whilst the crew of a collier when his father was a young man numbered at least ten, with a couple of apprentices on top, William considered that *Mary* was strongly manned with eight hands all told, of whom three might be apprentices. *Mary*, despite her employment carrying coals from Tyne, Wear and the South Wales ports to London, was not merely a coaster, and William was more than a shallow-water Master. In 1836, for example, after a couple of routine round trips between Shields and London, he took her by the northerly Viking passage to British North America. She left Shields on 14th June, arrived at Quebec two months later, loaded a cargo of pit-props and sailed again on October 12th. She arrived at Swansea for orders on 9th December and started to unload at Llanelli three days later. She was typical of her sturdy breed and age, immensely strong, powerful and handy. Heavily rigged, she crossed single topsails and carried a full suit of studdingsails. The clews of the fore-course were bent permanently to the newfangled Bentinck boom which became an almost universal feature of the brigs employed in the coal trade.

The brigs that William's father had sailed were, like George Fleming's *Jenny*, built with open deck rails aft and no protection above deck amidships; but in William's time the use of high quarter boards and bulwarks came into fashion. Always the bulk carriers of the north favoured the plain stem without the decorative headrails, trailboards, knees and figurehead so much beloved in other trades. During the French wars it was usual even for unarmed merchant shipping to show black ports along the topsides to deceive the enemy privateers, a custom which continued among the brigs and barques of the northeast throughout the 19th century, aided by cheap supplies of tar and 'black varnish' from the local gasworks.

Despite the industrial grime inherent in the business of loading and unloading coal, the bulwarks and inboard stanchions were painted brightly, usually green, white, pale blue or red. For firmness of footing, decks were either painted or tarred and sanded. At sea all was immaculate, for it was only contact with land that begrimed a vessel and corrupted the seamen.

Since smart anchor work in sluicing tides and confined waters played a vital part in their operations, the windlass was an important deck fitting, and it retained the simplicity of its early origins for a long time; an absence of mechanical complexity might not save labour but it did ensure reliability. William's son James saw the last years of the old simple barrel windlass and its replacement by the pump-handle 'Armstrong's Patent' which, whilst it certainly speeded up the shortening of the cable, never matched its predecessor's power to break an anchor out.

At sea the chain cable was always ranged along the deck, secured by lashings to ring bolts with enough for'ard of the windlass to enable at least one anchor to be let go quickly. When anchored normally the windlass bits took the strain of the chain cable directly; but when riding out a gale in, for example, the Downs or Yarmouth Roads, the load was taken off the bits by shackling a thick grass spring to the cable close to the hawse-hole and leading it aft to the two massive sampson posts, where it was heaved taut by tackles shortened on the cargo winch; the cable was then slacked off and the strain was absorbed by the spring as the bows rose and fell to the sea.

The high, narrow-based sail plan of the brig needed a bowsprit and jib-boom of relatively enormous length, fully one half of the ship's overall measurement, to carry the head-sails required for balance. Port regulations ruled that for safety's sake in crowded waterways, the jib-boom be run inboard above Greenwich or inside the Tyne bar.

The very first steamboat to appear on an English waterway was a little vessel, later called *Perseverance*, on the Tyne in 1814. She was intended as a ferry, but gradually picked up work towing sailing vessels, though the conservative colliery owners were slow to see the significance of this almost accidental evolution of the steam tug. It was the owners of the passage-boat services who quickly saw how steam towage could overcome many of the delays and hazards occasioned by the narrow entrances of the Tyne, Wear and Thames.

Eventually in the 1820s as more and more steamboat towage

became available, the coal factors began to see its value in liberating the colliers from the tyranny of the tides, particularly in the dark winter months when coal prices soared, with the long-term result that the coal trade was speeded up and the winter lay-up was finally abandoned.

Among the earliest of watermen to work in steamboats was Stephenson Chisholm, some eight years younger than William Johnston. The Chisholms were a sizeable clan in Shields and, like so many migrant Scottish families, they had cousins on Wearside as well. They became very much involved as Masters, Owners or Engineers in the new-fangled vessels, and consequently well-known to the Shipmasters and Mates of the port.

Joseph Fleming, having completed his apprenticeship in 1825, sought a berth as Second Mate. His beloved sister Mary's husband, George Hutchinson, who was at that time William Johnston's Mate in a Baltic Trader out of North Shields, was able to find a berth for the young Fleming in his vessel. On a winter voyage carrying coals to London, Joseph met and married the daughter of a retired mariner and former member of the Shields Scotch Church who was now keeping lodgings near the coal wharf, but in 1828 she died giving birth to his first son, John. Mary Hutchinson undertook to foster the child together with her own new-born daughter, Phyllis, at North Shields. This was the beginning of a lifelong and deeply affectionate relationship which ended only when Phyllis sat by John's deathbed in 1883 watching him cough his life out.

The Hutchinsons and the Johnstons were members of the Scotch congregation and so were the Chisholms. Stephenson Chisholm's son John was born in the same year as John Fleming and Phyllis, and the three of them grew up close friends. The young Chisholm became a steam-boatman in his turn and, advantaged by John Fleming's long absences as an apprentice in the Baltic trade, wooed and won Phyllis whom he married in 1850.

Joseph married for the second time in 1830, this time to the illiterate Hannah Thompson, daughter of a Bishopwearmouth postboy named Abel. They acquired a house on Monkwearmouth Shore, produced children in between voyages, and sometimes had John to visit them. But until he was old enough to go to sea, John, who disliked his step-mother, continued to make his home with the Hutchinsons at North Shields.

Joseph, after his second marriage, engaged as Mate in various snows including *Zephyr*, *Wensleydale*, *Moore*, *Pelion*, *Isis* and *Echo* all out of Sunderland carrying coals to London, Calais, Boulogne, Havre de Grace and Honfleur as well as down Channel to English ports. The snow was a variant of the brig. It had a small trysail mast close abaft the mainmast, resting on a step on deck and supported by chocks between the trestle trees of the maintop; this mast received the jaws of the gaff of the spanker and the luff of the sail hooped to it. The spanker of the snow, unlike the driver of the brig, had no boom.

In 1839 Joseph became Master of *Eliza*, for the next three years trading coal from Sunderland to Hamburg, general cargo thence to the Port of London, and sailing back to Sunderland in ballast. His brother Daniel, who had been *Eliza*'s Master prior to 1839, succeeded him when he took command of *Elizabeth Adnett*, a snow of 258 tons burden in which he sailed until 1851 in the coasting trade from the Tyne in the north, and down Channel as far west as Barnstaple, and to the continental ports between Bremen and Brest.

The development of the young John Fleming's career after serving his time was slow. He was still only a Second Mate when at the age of twenty-four he married Sarah Jane Lillie in 1850. She was offspring of a maritime family. The wedding feast was largely due to the initiative of his Aunt and foster-mother Mary Hutchinson, who made it an occasion for a gathering of the clans. Most of the wedding guests were women because the menfolk as usual were at sea, but Joseph Fleming came over from Monkwearmouth to see his first-born properly spliced.

For old times' sake William Johnston, who had been skipper to both Joseph Fleming and George Hutchinson, was invited; he and his wife Helen together with two unmarried daughters were living in lodgings at the Master Mariners' Asylum at Tynemouth, for his health had broken more or less completely.

Even by the time he was forty-five William had begun to show signs of failing, not so much in skill as in memory. His ability to perform ship's business ashore became defective and he was eased out of the Master's berth which he had occupied for over a decade in *Mary*. He was supported by his wife and two daughters working as dress-makers until his pride forced him to find a berth as Mate of a former London collier the *New Eagle*. She was a barque of nearly 300 tons, built in New Brunswick in 1825. Under Marshall, a good Master, Johnston performed his duties adequately on his lengthy but final voyage to the East Indies in 1843-4. Nevertheless the voyage finished him and he abandoned the sea for good. He died in 1861 after a decade of physical and mental decline.

James Johnston, the second child of William and Helen, was a gifted, handsome and ambitious youngster determined to succeed in the calling of his forefathers. When he was fifteen he was apprenticed in the *Content*, a London collier of the old school, a cat boat built in 1770. James learned his seamanship the hard way. In the middle of 1835 he transferred his indentures to Thomas Young and for the next four years he sailed under the command of Matthew Mather in *Mary Young*, a smart modern 240 ton Baltic trader.

His first voyage in the latter took him to Archangel. She dropped down the Tyne from Newcastle on 19th June 1835 filthy with a cargo of coal, but by the time the hands were sober and Captain Mather had taken departure from Buchan Ness every inch of her paint-work and canvas was sparkling. Fair winds carried her briskly beyond her landfall near Cap Stat, but she crossed the Arctic Circle slowly in a band of frustrating variables. Near North Cape the wind settled nor'easterly and the fog drifted down from the ice in the north. Once round the Cape, close-hauled on a southeasterly course, she soon broke clear into the summer sunshine again. The days grew surprisingly warm and James found it hard to believe that only six or seven weeks earlier the White Sea had been frozen up.

After unloading much of her coal into lighters, *Mary Young* crossed the Dvina bar with about 14 feet of water on it, berthed alongside in the harbour on the island of Sollenbole, about a mile from the old town of Archangel, and lay there for nearly a month offloading the rest of the coal and taking on board a cargo of timber which had been drifted down the Dvina. The river was some five miles wide here, and, being navigable by small craft for some 500 miles above Archangel, it was the main highway down which the port's produce was usually floated from the hinterland on large rafts. About half of the ships trading to Archangel were English; most of the rest were Norwegian. The old town, consisting of two principal, very irregular streets connected by narrow lanes and paved with wood, was ill-built. Most of the houses were wooden.

After she returned to Newcastle, *Mary Young* completed the year with three voyages between Tyne and Thames in the winter coal trade. During the early part of 1836 she voyaged first to Lisbon and then to Quebec. This was the pattern of the remaining years of James's apprenticeship—a protracted foreign voyage in the fair eight months of the year followed by three winter trips to London. His apprenticeship completed, he took a berth as Second Mate of the snow *Liddell* (328 tons burthen), built in Shields in 1761, trading to Bordeaux, and within two years was appointed her Master, coasting and trading to the north European ports. In 1846 he met Frances Livingstone, who had recently moved away from a life of rural poverty with her parents in Dumfriesshire to stay with relatives in Tynemouth, where she found work as a household servant and became another of Mary Hutchinson's warm circle of friends. After a rapid courtship these two strong characters were married in the same year.

William Livingstone, Frances's father, had been a rolling stone on land as much as the Johnstons were wanderers on the sea. He was a grandson of that William Livingstone who had married old Andrew Gibson's daughter Anne. Like his grandfather, his father John had been a mariner and, finding himself a wife in the port of Dublin, had made his home there. The sea, and particularly the prospect of forced service in the Navy, did not appeal to the young William, so in 1803 when he was eighteen he so far broke with family tradition as to enlist in the Army to go and fight Napoleon Bonaparte.

A few years later he was invalided out with a pension. In 1810 he married Rosanna Jones, the daughter of a small Irish farmer, and with her went to look for better working conditions and prospects as a handloom weaver in Dumfriesshire. Cottage weaving, however, was in rapid decline, so he took to the land and worked on various farms in the vicinity of Annan until he died at Holmhead in the parish of Half Morton in 1873. He was an amiable but somewhat restless workhorse of a man who sired a large progeny. Frances, born at Eastfield in the parish of Kirkpatrick Fleming in 1819, was his third surviving child.

The marriage of young James and Frances was an occasion for rejoicing; they were a handsome, strong and proud couple. Their first child William was born in 1848; Annabell followed in 1849; and James John, conceived in a fortnight's shore-leave between Baltic voyages, arrived in 1853. James's future seemed assured when, after three years as a shipmaster in the coasting trade, he became Master of the smart Baltic Trader, *Pomona,* in 1849.

She was a snow, built in Sunderland in 1836, of 284 tons burden and owned by Dixons and Co. For twelve years, as Master of *Pomona* and *Swallow* and *Vesper*, he followed the trade year by year to Stettin, a port of considerable stature for timber and corn; to Riga, which specialised in timber for masts and had been the principal hemp port until Cronstadt overtook it at the beginning of the 19th century; to Gothenberg; to Dantzig, Europe's principal grain port as well as a supplier of timber, potash and linen goods; to Memel which supplied grain and large quantities of flax; and Cronstadt, an important exporter of timber, linens and iron, where his family name was well known and respected; and to Narva, the greatest flax port.

James also made three spring voyages across the Western Ocean to North America. On 25th April 1851 he cleared Shields for Quebec with a cargo of iron. Because of the fog and ice on the Newfoundland banks at this time of year he took the southern route. *Pomona* lay at the Orleans Wharf for the first ten days of June loading timber for the voyage home. In September she cleared Shields again to catch a grain shipment at Cronstadt. She was late out of the Baltic that year and had a rough time of the winter storms.

The North Baltic ports being completely icebound during much of the winter, every ship tried to be out of it by November and no ship would sail thither before March. They had to be clear of the Gulf of

Finland and the Gulf of Riga in October and they often could not re-enter until late May. As well as the risk of damage from ice on passage, the high cost of paying and victualling crews kept idle in an icebound port made Masters take very cautious judgements about the prospects of the annual freeze-up. In so short a season the Baltic Traders usually achieved only two round voyages from England, the first sailing in April and landing a cargo back there in July, then making a quick turn-round to enable them to be back in England again at the end of November.

After returning from Cronstadt, *Pomona* filled in the rest of the winter of 1851-2 with the odd freight of coal to London or to the various ports between Brest and the Elbe—these two points defined the limits of what was called the Home Trade. In March '52 she made another round trip from Shields to Quebec, and after a seven day passage to the Pool of London she made a round trip to Riga, and was back in Shields early in January 1853. Next she was seven months away in the Baltic, returning to London in August, whence she made a round trip to Dantzig.

While James was on his way to the London River, his son James John was born in North Shields on 8th October 1853. James first saw him in December after bringing a cargo back from Hamburg. Then he became Master of the brig *Swallow* (249 tons), owned by the same people and built in Sunderland in 1834. That year his only deep-water venture was one trip to Quebec; and the next year he made one to Gothenberg; otherwise he steadily plied the Home Trade. Indeed, apart from one voyage to Stettin, one to Memel and his initial voyage in command of the brig *Vesper* (290 tons) to Cronstadt in '58, he remained in the Home Trade until the end of 1861 when he moved upwards into a 'South Spainer' barque, the newly built *Agenoria*

John Fleming's career was not dissimilar. Most of his early sea service was also in the Home and Baltic Trades in brigs. Apart from a voyage to Genoa in the *Hylton* in 1855 he sailed regularly into the Baltic until November 1860 when he took his Master's certificate at Sunderland.

He served as Mate in *Argonaut*, *Flying Childers* (a well-known

American clipper), *Jobling*, *Maybrough* and *Frankfort*. His first command was the brig *Elizabeth* in the Home Trade at the end of 1860, but in the following spring he took a berth as Mate in the South Spainer *Ambassador* carrying coals from Cardiff to Alexandria. A year later he joined the barque *Russell* (381 tons) at Shields as Mate for a voyage to Cronstadt, and on reaching Liverpool in June 1862 he was appointed Master. He remained so for the next five years.

BARK OR BARQUE

BARQUENTINE

BRIG

BRIGANTINE

FORE & AFT SCHOONER

9

The Barques

WHEN JAMES JOHNSTON BECAME A YOUNG MASTER IN the 1850s the construction of barques began to be fashionable again, particularly at Sunderland, but these were world traders of 400 tons or more which, although often freighted with coal outwards, were really ocean-going bulk carriers having little in common with the specialist colliers. The Masters and crews of these finer lined 'South Spainers' and the 'Baltic Traders', although they might seasonally engage in the London coal trade, held themselves to be several cuts above the colliermen in the nautical hierarchy.

In the mid 1850s old Joseph Fleming was still plodding along in the Baltic trade as Master of the snow *Mary Sophia* (265 tons) owned by his brother Daniel, but in 1857 he was fatally appointed Master of the barque *Severn* (330 tons) and started voyaging to South Spain. Having cleared from Shields for Carthagena in January 1859, *Severn* arrived back on the Tyne in June, sailed for Lisbon ten days later, thence carried a cargo of wine to Bristol. On October 20th she cleared that port to load coal at Cardiff for Rouen. Six days later she was lost on Hayle Bar in the great gale of that month which, it is recorded, caused the loss of 24 vessels.

The *West Briton* of 28th October reported nine lives lost in the wreck on Hayle beach:

> 'The barque *Severn*, 500 tons burthen, laden with coals, which sailed from Cardiff on Tuesday last bound for Havre de Grace, became a total wreck on Wednesday morning on Hayle beach. The ship's crew consisted of ten men. One seaman named John Taylor,

a native of Appledore, who only shipped on board the previous Saturday, saved his life by clinging to a plank, and washed ashore. It is reported that the Godrevy revolving light was mistaken for the Longships light. The vessel was taken in the late Russian war. The Hayle beach is covered with coals and pieces of wreck.'

In its next issue the *West Briton* reported that the night of the wreck was one of the most violent wind and heaviest seas for 36 years.

A vessel caught by a rising gale with the flood tide running against her at night off the iron coast of Cornwall was in terrible peril. There was nowhere to run for and no sea-room at all for scudding. The only course open to its Master was to heave to on port tack under deep-reefed main topsail with a prayer that she would fore-reach sufficiently to keep her offing.

No doubt after a hard day's beat down the approaches to the Bristol Channel Joseph had hoped to weather the Longships; but, having failed to pick it up, had gone about and hove to with some element of doubt about his precise position. He would not have had a sight of the coast for several hours. As the gale rose and *Severn* laboured more and more heavily in the high seas sweeping in with unimpeded force from the open ocean, he would be increasingly concerned that the overstrained topsails, deep-reefed though they were, would blow out of their bolt-ropes as the barque rose from the calm in the hollow of each sea into the full force of the gale on the next crest.

The rain and flying spume would have cut his range of vision to a few hundred yards. His heart must have leapt when, in a sudden, brief clearing, he glimpsed the loom of a revolving light on the starboard quarter. It could only be the Longships light, for it was exactly where he would have calculated it should be.

Now he could safely wear ship onto a broad reach to clear Longships to port and have all the sea-room he needed to ride out the gale. The tense, almost motionless watchfulness of the Master during the preceding hours while the barque had been lying to was transformed on the instant. "Stand by, Mister, to wear ship" he bellowed into the mate's ear. Then as she fell off in the next blast of wind and thrust of sea: "Helm a-weather, ease off handsomely on the mainsheet . . ."

He knew there was risk in wearing in such high seas under main topsail alone, but he judged that to try to set the fore topsail in the gale would only result in its being blown to tatters in the act. He stood close to the man at the wheel to see that he did not let her head pay off too quickly without gathering headway as he ordered the yards to be squared step by step.

By the time the wind had come abaft the rain was thick again and there was no sign of the light. Secure in the knowledge that he had a well-found ship and plenty of sea-room whatever wind and wave might do, Joseph went below, leaving the Mate on watch with his sailing orders.

The tragedy was that either he did not know, or had forgotten, that only a matter of weeks earlier a new revolving light had been set up on Godrevy Island to mark a notorious death-trap for shipping. Instead of heading for the open sea, *Severn* was now driving straight into the shoal waters of St Ives Bay.

Once she was embayed, given a gale from this quarter and such sweeping seas, no brilliance of seamanship, could save her. The end must have been quick. In pitch darkness, the scream of the gale in the rigging deafening the watch on deck to the noise of the surf ahead, the first indication of danger would have been a sudden steepening and breaking of the seas, probably causing *Severn* to broach to, and by the time Joseph reached the deck she would have been totally out of control with one or more masts gone over the side. In a few more minutes she would have struck the sands.

Joseph Fleming and all his crew but John Taylor were drowned.

During the 1860s both James Johnston in *Agenoria* and John Fleming in *Russell* found themselves heavily engaged in the Black Sea grain trade which had opened up after the Crimean War. This trade formed a profitable partner to the outbound carriage of bunker coal for the rapidly growing steamship traffic through the Mediterranean. *Agenoria* was a barque of 377 tons register built in Sunderland and launched in 1862. She was owned by J. Storey of Gateshead and J. Urwin of Newcastle. They knew how to manage their ship and

Johnston, her first and only Master, knew how to sail her.

Between March 1862 and February 1867 she made eleven voyages to the Mediterranean and Black Seas, each of about seven months duration between clearing outbound from a UK port to first arrival inbound. Over the first seven full years of her life she spent less than seven months in UK waters either in port or in transit between home ports. During virtually the same period John Fleming was Master of the two barques called *Russell*, both about 380 tons gross register and built by Davison of Sunderland for W. Atkinson.

The first *Russell*, 121 feet long, 27 feet beam, 18 feet depth of hold, was launched in 1859, the second in 1867. Between November 1861 and December 1868 John Fleming made ten voyages, each of about six months duration, six of them to the Mediterranean and Black Seas. That the *Russells* were less well managed than *Agenoria* is evident from the fact that in roughly the same seven years the *Russells* were idle between voyages for about 24 months in UK waters.

Agenoria's voyage to the Black Sea in 1864, although of longer duration than usual, does illustrate the sort of passages she was making. Having unloaded a cargo of grain from the Mediterranean at Glasgow in the middle of January, she sailed north about to the Tyne to load coal for Venice. She cleared outbound on February 24th, proceeded via the Downs, beat down Channel and, in order to avoid any risk of being embayed in Biscay or having the Portuguese coast too close under her lee in a season of gales, made a long board to gain an offing in about 12° west longitude before pointing free for Cape St Vincent and Gibraltar. Her subsequent movements were reported thus:

15th April: arrived Venice
6th May: sailed from Venice
27th May: Constantinople
1 June: Kertch
4 June: arrived Taganrog
29th June: sailed from Taganrog
5th July: Kertch
23rd July: Constantinople

26th July: Dardanelles
18th September: arrived Falmouth for orders
26th September: sailed from Falmouth
4th October: cleared inwards at Glasgow.

The passage between Venice and Taganrog was hard work for all hands, for the holds, filthy from coal, had to be thoroughly cleaned to be ready for wheat and then covered all over, floors and sides, with bamboo mats. The mats were supplied to the charterer of the ship by the master and traditionally were the source of some small profit to him.

At this stage the carpenter came into his own, for it was his task to make and supervise the fitting of shifting boards and linings and box feeders. At that time, bulk grain was one of the most dangerous cargoes to be carried, because of the ever-present risk of its shifting; and apart from the complex Board of Trade regulations which had to be complied with, the know-how of the carpenter and the meticulous supervision of the Mate during loading were vital to the safety of the ship.

After this voyage *Agenoria*, having discharged her cargo at Glasgow, sailed round to the Tyne, loaded coal, and cleared outward to Constantinople on November 11th.

John Fleming's first four voyages in *Russell* were to the Baltic. She then lay in Liverpool for a couple of months until 14th December 1864, when she cleared outwards for Buenos Aires. She arrived on 7th February and sailed on 20th, fetching the chops of the Channel in mid July and berthing in Antwerp on 23rd. Thence she proceeded to Cronstadt (29th August), anchored off Hull on 14th October, unloaded at Grimsby and coasted in ballast to Shields. In November she left for Odessa. Two Black Sea voyages followed, the second of which terminated at Antwerp in February 1867. There she picked up a freight for Havana which included two iron boilers upwards of ten tons each as deck cargo. She sailed on 13th April.

Whilst beating through the narrows of the Channel she encountered a heavy squall that put her almost on her beam ends, causing the two boilers on deck to break adrift and smash three boats, start the deck and main hatchway, and do much minor damage besides. Fleming hailed a tug to take her into Dover, but the tow parted, so he decided to take a fair wind to the Downs. There she lay for several days while he directed the Carpenter and hands to carry out the necessary repairs. By May 7th she was ready to sail.

Clearing out of Dover for Havana on 9th, *Russell* beat down Channel and made the customary long board to get the best offing to westward of the Bay of Biscay. When Fleming reached longitude 12° west he estimated that he would give Finisterre a wide enough berth on starboard tack, and shaped course for Madeira. After passing the island he shaped course to cross the parallel of 20° north somewhere between 25° and 30° west so as to catch the northeast trade wind as soon as possible, then ran down his westing along the latitude of the channel between Santo Domingo and the tail of the Bahama coral banks to the north.

It was not the safest approach to the Old Bahama Channel, for it was the least well marked. On 24th June *Russell* ran smack into a steep-to coral bank of the Plata shoals, part of what is now known as the Silver Bank to the north of Santo Domingo. She became a total loss. John Fleming and his crew, having taken to the boats, were picked up by a passing steamer and landed at Havana on 30th.

Russell's owner, Mr Atkinson, rapidly acquired a second *Russell* with the help of insurance money, and just over two months after the loss of the first one John Fleming cleared her outbound from Shields with coals for Constantinople. After a five week passage of nearly 3,000 miles borne mainly on favourable winds, *Russell* sailed past low-lying Cape Matapan, the uninspiring most southerly point of Greece, raised the bold, mountainous headland of Cape Malea in daylight, and altered course into the crowded Aegean Sea.

John decided to take the usual channel between Elafonosis Island and Kithera, some four and a half miles wide, despite the likelihood at that time of the year that the weather might change very rapidly and bring on strong blows from unexpected directions.

And so it was. Through the Elafonosis Channel and up through the Aegean Sea the winds were fickle and sometimes fierce, and it took

Captain James Johnston (1819 – 1872)
and his bride, Frances Livingstone (1819 – 1893)

another couple of weeks of tiresome sailing to reach Constantinople only 400 miles further on. In March *Russell* ended up in Marseilles. No further freight offering, her crew was discharged. John travelled home overland. It turned out to be his last Mediterranean voyage.

James Johnston meanwhile continued in the Mediterranean and Black Sea trade. In January 1873, laden with wheat from the Ukraine, *Agenoria* called at Falmouth for orders. Proceeding thence to Dublin she encountered strong westerlies once round the Lizard, with thick sea fog and a heavy swell coming in from the Atlantic.

For whatever reason, James failed to make sufficient offing to westward and drove her onto the Longships reef. When the fog lifted at noon on 26th January her topmasts were seen sticking out of the water, but the tremendous groundswell prevented the approach of a coastguard boat sufficiently close to identify her before she broke up.

James and his crew of ten perished. On 3rd February the Shields Daily News reported:

> 'From the accounts received here there is too much reason to fear the ship *Agenoria* of Newcastle has foundered. In the course of the last week it was reported that a jar marked *Agenoria* had been found washed up near Falmouth, and that some wheat was found floating about a wreck. A later account sent from Penzance on Saturday states that the following articles have been picked up from the barque sunk near Longships: part of a topsail marked 'Harrison Sailmaker, Newcastle', also a card of bonded stores addressed 'HM Customs, Sunderland from J. Snowden, 51 the Quay, Newcastle, *Agenoria*. A quantity of wheat is also washing about. The *Agenoria* was laden with wheat and left Falmouth under orders for Dublin. There was no account of the crew, and the worst forebodings are entertained regarding them. She was commanded by Captain J. Johnston of North Shields, and sailed from Sunderland, where the crew was shipped. The *Agenoria* was the property of Mr J. Storey of Gateshead and Mr J. Urwin of Newcastle. She was 376 tons register and was built on the Wear in 1862.'

James' widow Frances lived for another thirty years, falling overnight from the respectable affluence of a successful shipmaster's wife to the hardship and grind of earning by her own endeavours enough to bring up five children (the oldest son William and James John were already supporting themselves at sea).

She was proud and she had steel in her. When the children had grown up and she saw them earning in respectable trades she went to live out her last years with her parents in Dumfriesshire.

In this tale of men who lived on the sea, their womenfolk are shadowy creatures like the one who appears in the burial register of Inverkeithing Parish on 9th December 1825 as simply 'William Johnston's wife.' In the small burial ground of the Kirk at Kirkpatrick Fleming, looking over the Solway Firth towards Criffel and Skiddaw, there is an eroded gravestone which reads:

> 'William Livingstone who died at Holmhead, Half Morton, 4th March 1873 aged 89 years. Also Rosanna Jones, spouse, who died 22nd May 1874 aged 85 years. Also Anne their daughter who died 9th July 1843 aged 17. Also their daughter Frances, relict of Captain James Johnston of North Shields who died 14th July 1893 aged 73. Also Agnes Jones, Mother of the above Rosanna Jones who died May 1844.'

That phrase 'Frances, relict of Captain James Johnston' sounds like a proud trumpet call from the grave. Fanny was a strong woman who bore her husband's children as they were conceived regularly in the short intervals between long voyages, held the family together and kept a respectable home for a husband who was rarely there, and made the difficult adjustments that were necessary when he was there, and remembered him faithfully during the long years of her widowhood. A flash of her metal is perpetuated in that single phrase on the gravestone.

After his Mediterranean voyage in the second *Russell* John Fleming made a couple of voyages in 1868 as Master of the *Try Again*, and then fell upon hard times. It was as though the ordeals of the ill-fated Havana voyage had sapped his nerve and his will to command even in such familiar waters as the Baltic. For a couple of years he made do as a ships' husband.

He had a cousin John Fleming, some four years younger than himself, now Mate of a coasting steamship *Seaton* out of Sunderland, who influenced him into believing that being a watch-keeping officer in a steamer was not too arduous a way of earning a living. At the age of forty-two he was persuaded to take a berth as second mate in the newly launched 1200 ton steamship *Nelson* owned by Harry Nelson of Westoe. Her voyage to Cronstadt and back to London, her port of registry, occupied 33 days as compared with nearly three months under sail in *Try Again*. Despite his loss of status he must have discovered some virtue in a steamship, for, though he was engaged to join the 1,000 ton full-rigged ship *Savely Chloudoff* at Cronstadt in the summer of 1872, he never went to sea in sail again.

After *Nelson* he obtained a berth as First Mate for a short voyage to Ferrol in a steam collier *J.A. Gibb*. Next he shipped as Second Mate in a new 1,000 ton iron screw steamer *Hotspur* for a couple of Mediterranean voyages during which he was elevated to First Mate. His last voyage was as First Mate on the maiden voyage of a 1200 ton steamship *Ben Lomond* to Odessa and back: voyage time two months and ten days compared with six or seven months in *Russell*. And that ended his seagoing career at the end of 1872 at the age of forty-four.

He lived for another eleven years in the Master Mariners' Asylum at Tynemouth before dying of exhaustion after three years of severe bronchiectasis.

PART THREE

Clyde Sail and the birth of the Steam Ship

10

Sail On The Western Ocean

UNTIL THE UNION OF THE KINGDOMS OF ENGLAND AND Scotland in 1707 the English Navigation Laws denied the Scots any part in carrying people or goods to or from the English possessions along the western seaboard of the North Atlantic. The ports of London and Bristol therefore dominated the trade on the Western Ocean.

Always disadvantaged by the growing ascendancy of the English Merchant Marine over all comers, Scottish shipping from her western ports on the Clyde and the Solway was constrained to the coastal trade, to a small way of trade with Europe, and to the fisheries. In the 15th century Glasgow was trading smoked salmon and herrings to France in exchange for brandy and salt. By the 1650s she was trading to Ireland small coals in open boats from five to ten tons in exchange for hoops, rings, barrel staves, meal, oats and butter; to Norway for timber, and to France for her wine.

There had even been occasional ventures to the West Indies, but at such loss as to discourage any thoughts of rivalry with the English. Any mercantile urge the Glaswegians may have had was held severely in check by the shallows of the River Clyde which prevented vessels of burden coming up nearer the town than fourteen miles, where they had to transfer their cargo into small boats of some five tons displacement known as 'gabarts'.

In 1651 there were twelve vessels belonging to the merchants of Glasgow, three of 150 tons each, one of 140 tons and the rest under 50 tons, none of which could go up to the town. Not long after this the Bailies of Glasgow resolved to build their own harbour near the head of navigation of the Firth and bought twenty-two acres of undeveloped land on the Bay of Newark—later to become Port Glasgow.

Until the beginning of the 17th century Greenock had been a mean fishing village of a single row of cottages, but the grant of a Charter by Charles I to John Shaw to erect the village into a free Burgh of Barony led the town into acquiring some ships that engaged in the coasting trade and, in a small way, the foreign trade also. Its herring fishery grew to employ some 900 boats, each crewed by four men and setting up to twenty-four nets at once. Immense quantities of these fish were exported to France, Sweden and the Baltic. In 1674 a Company was formed in Glasgow to enter the whaling industry and make soap; it had five ships and extensive premises in Greenock for boiling blubber.

Inspired by the colonial enterprises of the English, the Scots determined to become a trading nation in their own right. In that period of tranquility following the abdication of King James VII and II they entered enthusiastically into a scheme to colonise the Darien Isthmus—the region now known as the Isthmus of Panama. The citizens of Glasgow led all Scotland to subscribe with alacrity most of their means to the scheme, and many of them embarked on the several expeditions to take possession of a thinly populated tropical wilderness and, centuries before the Panama Canal was constructed, to control trade between the Atlantic and the Pacific Oceans. The last, four frigates and 1200 emigrants, sailed from Rothesay in September 1699.

This great national undertaking failed disastrously, partly because of the jealousy of the English and the obstructiveness of King William, and partly because of the open hostility of the Spaniards. Out of all the great number of hopeful and energetic emigrants who sailed westward from Forth and Clyde, little more than a score or two of broken and beggared men ever saw Scotland again.

At home hundreds of affluent families were ruined by the loss of their investments. Glasgow itself suffered so severely that another fifteen years were to pass before any of Glasgow's merchants could afford to possess ships of their own. A great mercantile dream had become a nightmare.

The one measure which had been regarded as the probable ruin of Scotland, namely the Act of Union with England, actually gave

Glasgow its first real stimulus. Even though the traders on the Clyde were forced to charter English ships, mostly from Whitehaven, the Union opened up to them a free commerce with the American colonies and the West Indies which had hitherto been denied. In 1710 Port Glasgow was constituted as the principal Customs port of the Clyde, with Greenock as a Customs branch of it.

The Glasgow merchants sent out their ventures in chartered Whitehaven vessels to Virginia and Maryland to exchange for tobacco. It was their custom to send out a supercargo with each vessel who disposed of the goods with one hand while acquiring the tobacco with the other. Credit was neither asked nor given. This primitive trade so prospered that the first Glagow-owned vessel in this trade crossed the Atlantic in 1718. Built in Greenock, she registered only 60 tons. The first Greenock-owned vessel in the trade sailed the following year. The Glasgow merchants quickly secured the lion's share of the tobacco re-export trade and even began to undercut the English in their own home market.

The halcyon age of the tobacco trade, during which the Glaswegians made prodigious fortunes, was between the mid 1730s and the outbreak of the American War of Independence. The greater part of the disposable capital of Glasgow was invested in it. The number of ships, brigantines and sloops belonging to Glasgow in mid-century was 67, trading to Virginia, Jamaica, Antigua, St Kitts, Barbados, Gibraltar, Holland, Sweden and Ireland, besides maintaining a substantial coasting trade.

The sailing tracks from the Norse waters and the Baltic, from the English Channel and from the Mediterranean Sea to all the oceans of the world run through the North Atlantic. This, the 'Western Ocean' of the old sailors, bred its own special type of Masters and Mates. Although it did not share the awful majesty of the Southern Ocean, its structure of winds, currents and weather gave it a diverse character which sometimes could be just as vicious.

In the 1860s, for example, the full-rigged iron ship *Severn* in which Friedrich Linck served as Mate ran her easting down through the tumultuous greybeards and among the vast ice-islands of the Southern

Ocean's Roaring Forties, and survived a cyclone in the Tasman Sea, only to founder through stress of weather barely a couple of hundred miles sou'west of Bantry Bay.

Tobacco was a seasonal crop, but stripping and packing it for shipment went on for several months. As the century advanced, moreover, the plantations moved further and further back into the regions beyond the tidewater. This led to the development of the practice of warehousing in the colonies. Consequently, although there was a peak of sailings from the Clyde in the late spring and early summer, vessels usually arrived back there throughout the following winter and early spring because of the long delays in loading up the creeks of the Chesapeake. The crossing of the Western Ocean was not a long one—eight weeks more or less, either way—but the ships that took part in the Virginia trade rarely made more than one voyage a year.

Two shipmasters engaged prosperously in this trade were William Harvey, grandfather of John Harvey of Alloa, and James Watson, who was Master of the ship *Grinock* of Port Glasgow. They would leave Port Glasgow early in May and arrive in the Chesapeake in mid-July. They presented themselves for loading at many points up the rivers and creeks and indeed, unless crops were particularly heavy, were often forced to tout for business to complete loading.

The customary layday allowance in Virginia Charter Parties was between 80 and 100 working days. Vessels would not clear Cape Charles outbound until October or November, to arrive back early in January at Port Glasgow. There the tobacco would be unloaded three or four tons at a time into gabarts, to be floated over the Clyde shallows to the Glasgow warehouses.

Preparation of the ship, repairs and refitting for the next annual voyage barely allowed time for the occasional trans-shipment of some of the tobacco to a French or Spanish port and the collection of some foreign goods towards the cargo for Virginia. The outward freight was, however, not very considerable: most of the ship's earnings came from the homeward freight of tobacco, but from early days there was a regular carriage of passengers which added to the owner's earnings and, to the extent that the Master was responsible for victualling, helped enrich him too.

In the 1750s there was a popular Innkeeper in Port Glasgow named William Fleeming. During his lifetime he was recorded progressively in the Parish Register as a 'workman in Port Glasgow', Flax-dresser, an Inn-keeper, and a Merchant. Under his roof a close community of people met habitually, including sea-captains like James Watson and William Harvey, workers such as Robert Leckie in the manifold trades supporting the shipping industry, and the minor merchants of the port such as the Robertsons and the Pollocks.

They met there as much to do business as to drink socially together, and the families not only shared with Fleeming the prosperity of the rapidly growing maritime trade of the port, but also became linked by a series of marriages. Fleeming himself married Isabel, the daughter of Alexander McLeish, a gabartman who was a regular at the inn.

Captain James Watson's sailor son William married Isobelle, a younger daughter of Will Fleeming who was then described as 'late Grocer', the appellation probably meaning that he was a wholesale merchant in sugar, the commodity which founded the second wave of prosperity for Clyde shipping.

James Watson's great-grand-daughter Margaret Leckie Watson was to marry Captain William Harvey's great-grandson Frank, and their daughter Charlotte came to marry the great grandson of Will Fleeming's other daughter Agnes.

As Greenock and Port Glasgow grew in importance, so did the demand to enlarge their harbours. The severe limitations of shipping bulk cargoes in gabarts also created a demand to deepen the Upper Clyde, and in 1773 the brilliant young James Watt, son of a small merchant in Greenock, was commissioned to draw up plans for this. He had already made a local reputation for himself as a civil engineer through his survey for a canal between Forth and Clyde which failed to obtain Parliamentary sanction and, among others, his surveys of the lines of the Crinan and the Caledonian Canals.

It was, of course, in the 60s when, working as the mathematical instrument maker at Glasgow University, the task of repairing a model of the Newcomen Steam Pumping Engine set his mind to improving its low efficiency and led directly to his invention of the steam condensing engine which subsequently became the basis of the commercial viability of the steamboat.

Despite his maritime connections, Watt showed little interest in this application of his invention, and it was left to others to modify his double-beam engine into the side-lever engine that was to be the first standard marine prime mover.

Glasgow had had some little intercourse with the West Indies since the 1730s supplying domestic necessaries to the Planters in exchange for part of their sugar crops, but generally the trade was of minor importance until the revolt of the American colonists in 1775 ended the monopoly of the tobacco trade. The Glasgow Merchants were quick to expand the West Indian sugar trade to take the place of the tobacco they had lost. Moreover, while the English were heavily embroiled in the wars the Scots took the opportunity to open up commercial relations with other parts of the world, the produce of which they had hitherto received only at second hand. The virility of the merchants of Glasgow and the Clyde at this period is in strong contrast with the contemporary decay of the trade of the Forth ports.

Isobelle Fleeming's eldest son William Watson, who was born in Port Glasgow in 1790, worked there as an accountant's clerk alongside two budding sugar merchants, George and Robert Leckie, whose grandfather was Will Fleeming's friend and contemporary Robert Leckie. George Leckie took a chance to settle in Demerara when it became part of the newly created colony of British Guiana after the end of the Napoleonic Wars.

He did well and, in 1828, formalised a partnership with brother Robert, who had settled in Greenock, the sugar refining centre on the Clyde. The firm traded in Demerara as Leckie, Murphy and Coy., and in Greenock as Robert Leckie and Coy., carrying on a commission and mercantile business between them.

But the appalling climate in British Guiana proved too much for George and he died there in 1834.

Port Glasgow then was already suffering a severe decline as a trading port. When he was twenty-three years old, William Watson the accountant married the Leckies' sister Margaret, and in 1840 the couple eventually followed Robert to Greenock. Three years later Captain William Harvey's grandson Francis—'Frank'—sailed into the Clyde in command of the *William Mitchell* of Alloa, stepped ashore to visit some of his family connections, married Margaret, the daughter of William and Margaret Watson, and settled for the rest of his life in Greenock. A generation later, by another of Fate's turns, Charlotte, one of his daughters, was to fall in love with and unwisely marry John Wilson, a great-grandson of Agnes Fleeming.

Throughout the decade from 1843 Captain Frank Harvey was Master of the Glasgow registered barque *Ellerslie* (743 tons), owned by Messrs Handasyde and Henderson of Glasgow, who in their earlier days had been engaged chiefly in the Mediterranean trade. She was a wooden vessel built in 1841 at St John's, New Brunswick, during that decade when the North American ship-building industry initiated the final development of the sailing ship, culminating in the superb clipper masterpieces of Donald McKay.

Ellerslie, however, was the fruit of an earlier revolution: the Americans not only built ships cleverly and cheaply in softwood, but also studied the basic principles of wind-propelled vessels in a way never before undertaken. In the first half of the 19th century they contributed as much to the development of the sailing ship as the rest of the world had contributed in three centuries.

They went for speed by lengthening the hull in proportion to the beam, giving their vessels finer lines, a greater spread of sail and larger carrying capacity. By comparison with Captain James Johnston's traditional, sturdy Baltic Trader *Pomona* which lay near her in the basin at Quebec in the summer of 1851, *Ellerslie* with twice the capacity, looked, and was, a greyhound of a ship. She was typical of that special breed, the North Atlantic Packet ship.

Between 1843 and 1845 *Ellerslie* made two voyages a year between the Clyde and Quebec carrying mixed cargoes westward and timber back to the Clyde shipyards. In the spring of 1845 Frank Harvey entered into partnership with his wife's younger brother James

Watson in a chandlery business in Greenock. That was the year of the great potato famine in Ireland and the beginning of a gigantic flow of emigrants through Liverpool to North America. In the following decade over two and a half million people sailed from British ports in the unprecedented migration to America, and of course *Ellerslie* became involved in the traffic.

From then on she sailed more often from the Mersey than from the Clyde, and increasingly into the United States ports where the main tide of immigration flowed. On her return voyages cotton became her main freight.

In April 1851 *Ellerslie* sailed from the Mersey to Quebec, where she lay for ten days alongside the Orleans Wharf loading timber for the return voyage at the same time as James Johnston's *Pomona*. Fate decreed that one of Harvey's great grand-daughters was to become the wife of Johnston's great-grandson, the author of this book.

Four months later *Ellerslie* cleared out of London for Quebec again. In October she left the Mersey for New Orleans and arrived back on the Clyde with cotton in May 1852. After three weeks in her home port she voyaged to New York and arrived back in Port Glasgow in October. Three more weeks in her home port, then she made a four-month voyage to New Orleans which terminated on the Mersey in March. In April she set out on a three-month round trip to Miramichi (New Brunswick) for timber which she took to Dublin, whence she immediately departed on a five-month voyage to Quebec which brought her back to Liverpool on Christmas Eve 1854.

By the time Frank Harvey reached the Clyde in January he had decided that a long spell ashore was due to him. He wanted to enjoy family life with his three children, Frank, Meg and the enchanting little Charlotte. The substantial house at 20 West Stewart Street was inviting after so many years at sea encompassing only the mahogany and brass appointments of the captain's cabin and his ten or so paces along the weather rail of the poop deck.

After all, he was now comfortably off at the age of forty, a good time to establish himself as a personality in that society which his wife graced with such dignity. And the chandlery business in Rue-end

Street, which he had started up in partnership with Jim Watson, could do with some of his attention to stiffen it. So for a while he became a man of business in Greenock.

He moved easily enough through the world of shipowners as well as the masters of the ocean-going fleets, and made his mark with shipbuilders both large and small, and rapidly enhanced both the standing and the turnover of the firm.

And at home, in due course, his wife Margaret produced for him a third daughter, Isabella, to be known as 'Bee'.

11

Round Cape Horn

IN FRANK HARVEY'S SECOND YEAR ASHORE HE FOUND that life was beginning to pall. Among his customers were Handasyde & Henderson, who were now successfully establishing a major fleet of vessels. In 1852 they had merged with the Anchor Line, which was one of those set up on the Clyde in the 1840s to run steamships regularly between Glasgow and New York.

By the time Frank Harvey died their interests would have included the Valparaiso packet sailing ships, the steam packets which replaced *Ellerslie* and her sisters on the run between Glasgow and Quebec/Montreal, Glasgow-Lisbon steam packets, Mediterranean and Levant steamships, and the Portland and New York steamers.

The Company interested Harvey both practically and financially in a scheme to build at Balfour's yard in Methil, Fife, a Valparaiso packet to be constructed of iron, about 450 tons gross, and ship-rigged. Harvey not only financed part of her, both personally and through the chandlery business, but also undertook to supervise the building and fitting out of the vessel which was to be named *Dalhousie*. He and Jim Watson also agreed that here there might be a profitable opportunity to expand their own business towards the South Pacific. Valparaiso, which had been growing since the 1840s at a rate exceeded only by San Francisco, was the first port of any significance, a general rendezvous of vessels entering and trading in the South Pacific. Lying opposite Australia and on the direct route between the eastern seaboard of USA and California, its already substantial cereal trade boomed when the Gold Rushes took place in these two countries. As the port of the capital of Chile it was also the natural outlet of a broad and fertile hinterland.

Although Lloyds surveyed their first iron vessel, a ketch, in 1836, iron construction was in a relatively experimental stage until about 1840. Shipwrights knew little or nothing about the properties of iron and the way to fashion it to make a ship, nor did they desire to learn. Whilst the smiths, who did know how to handle the material, had none of the skills to design a vessel that operated in a fluid medium alien to their experience. So it was that the civil and mechanical engineers were gradually drawn into the business of shipmaking.

At the end of the 40s and the beginning of the 50s some shipyards, chiefly on the Thames, Mersey, Tyne and Clyde, were working entirely on iron vessels, particularly paddle steamers, and many were building in both iron and wood. By the time Harvey came to be associated with the building of *Dalhousie*, the use of iron in the form of knees, beams, hold pillars and diagonal trussing to strengthen wooden ships, was beginning to mellow the traditional wooden builders' old hatred of the metal.

The arguments in favour of an iron ship which chiefly appealed to Harvey and his associates were that, in the early 50s, the first cost of an iron ship (around £14 a ton) was lower than that of a wooden ship of the same class; that the carrying capacity was much greater for the same tonnage; that the material was immeasurably more durable; and that the running maintenance was much cheaper.

Before 1850 most of the iron sailing vessels were schooners, brigs and barques of small tonnage, but after 1852 the construction of full-rigged iron ships of substantial size became the standard pattern, and the building of small sailing vessels became rare. In March 1853 there were thirteen iron sailing ships building on the Clyde, twenty-six iron paddle steamers, fifty-two iron screw steamers, six wooden sailing ships and one wooden screw steamer. Fourteen of the iron vessels were for the expanding Australia trade.

Nevertheless when Lloyds' earliest survey rules for iron ships were published in the 1855 Register Book for the first time, they were prefaced by the remark 'Considering that iron ship-building is yet in its infancy, and that there are no well-understood general rules for building Iron Ships . . .' Harvey was therefore in the van of progress and taking no inconsiderable risk in the *Dalhousie* venture: there were still few precedents on which he and his colleagues could base their opinions, and the development of the new technology and its application to vessels of ever-increasing size had to be tested in the hard school of commerce before it could be judged a success or a failure.

Although she had a curved clipper bow in the Aberdeen fashion and a fine run to her counter, *Dalhousie* was an ugly vessel in her time, slab-sided, single-decked and totally lacking in sheer. Save for the white bulwarks, her topsides were black. But she looked what she was, a rugged vessel to fight the fiercest seas in the Southern Ocean.

As her building progressed, it became the obvious and inevitable conclusion that Harvey should sail as her Master. Launched in the autumn of 1856, she sailed under his command round to the Clyde. Whilst he shared Jim Watson's enthusiasm for the opportunity to seek business for their chandlery firm in Valparaiso, the real attraction of this new venture was the prospect of voyaging by way of Cape Horn, a fitting challenge even to a Master who had learned his trade in the Roaring Forties running down his easting to Australia and had passed his middle years bending the boisterous winds of the Western Ocean to his will to make fast and regular passages.

Many a night he spent poring over chart and log with other Captains of the line. He was to sail in November and arrive at Valparaiso some three months later at the height of the southern summer. As far as the southern limit of the southeast trade wind, at that season well to the south of Rio de Janeiro, he would be following a route already well known to him, standing across the wind on the port tack with no hazard but the steep-to rock of St Paul in mid-ocean.

It was usual to pass some 200 miles east of Rio. The older navigators argued that between October and January when the nor'easterly winds prevailed and the current set to the sou'west, it was best to close towards the Patagonian coast within 100 miles so as to avoid the heavy seas which the occasional westerly gales raised further offshore, and to profit by the variability of the inshore winds. They also argued that as the winds off Cape Horn were almost invariably westerly at this time of year, rendering the making of westing particularly difficult, the best passage was through the Le Maire Strait, providing a valuable saving of sixty miles.

It was a somewhat fraught passage, best attempted only in daylight with fair wind and tide, for the tidal streams are strong and the seas are viciously turbulent on a weather-going tide. There was, on the other hand, a growing school of Masters who argued that the stout, well-found vessels of the day were advantaged by standing a couple of hundred miles or more off the Patagonian coast and taking the almost direct route through about 45° south, 60° west, to pass 30 or 40 miles east of Staten Island before altering course to pass some 100 miles south of the Horn.

They made the point that there was no advantage, even with a fair wind and tide, in going close to the Horn, for close under it the east-running Southern Ocean current sets fast. After passing east of Staten Island a vessel should, however fierce the sea, they argued, be kept on starboard tack until in 60° south if the wind were westerly, unless it veered to the south of SSW, and then onto the tack that gave the most westing.

Both in the wider strategic sense of routing via the Strait of Le Maire or to the east of Staten Island, and in the tactical sense of reacting to local wind-shifts, the winning of every mile of westing was of primary importance. Whatever the routing and whatever the winds, the Master could be certain of one constant factor: the huge seas. In terms of skill, Harvey was at the top of his profession: he was now to discover that the despatch with which, voyage after voyage, a vessel could safely negotiate the Horn westbound depended on his own determination, persistence and unflinching courage.

But in the grey, damp dawn of 16th November 1856 he had more prosaic and squalid things on his mind as he stepped on board *Dalhousie* lying at the Tail o' the Bank. The impatient tug, belching black smoke from her high stack, was already hauling the tow-rope aboard, and her paddles churned white water in small bursts as her engines turned to keep her on station. Already Harvey's two Mates were beginning by sheer force of personality as well as many a blow to knock the mob of half-drunken seamen into a crew. Out of the chaos of the rabble, some men of sterner stuff and experience began to emerge, to make their mark on those of lesser breed; the two Mates marked them instinctively, and used them.

At Harvey's order, relayed by the first Mate, *Dalhousie* was hove short on the anchor and then broken out when the tug took up the strain. The first of the ebb was just beginning to run. All day the ship followed the tug down the Firth of Clyde while the crew cleaned her and made her shipshape. Running rigging, boats and equipment were cleared, checked and inspected; the potash swabs and deck brushes and holystones were put to work to take away the landsmen's filth, and in the grey of the afternoon the Mates, picking the men alternately, chose their two watches.

So quickly were the rough edges rubbed off that a real crew was beginning to take shape that night when Harvey dropped the pilot and tug off Rathlin Island. The tug bustled off with three cheerful blasts of her whistle, and under a clear sky and on a calm sea *Dalhousie* cleared out of the North Channel, close-hauled on port tack with a steady breeze coming off the land. The chains were unshackled from the anchors, the hawse-holes sealed and the two great bower anchors secured to the cat-head.

Sharp on eight bells of the second dog-watch, watchkeeping routine was set. The voyage had really begun: *Dalhousie* leaned to the wind, her forefoot cutting smoothly through the water, the sea gurgling softly in the wake of her sternpost, the wind humming in the rigging and drumming gently on the sailcloth; the real life of seafaring began all over again, binding ship and men to each other in the rhythm of the changing watches punctuated every half-hour by the ringing of the ship's bell.

Once out of the North Channel, Harvey's main concern was to make sufficient westing to be able to weather Cape Finisterre on the long reach for Madeira and the trade winds. As soon as these were picked up, the heavy weather sails were sent down and fair weather ones set in their place. His next problem was how to pass through the doldrums between the northeast and the southeast trades with the least delay. For both Master and crew the fine weather in the constant breeze and the calm seas of the trades, with *Dalhousie* booming along steadily day and night, brought about something almost approaching contentment. Fishing lines were put out for tunny, dorado and bonito; the adventurous scrambled down onto the bobstay to try to harpoon the escorting porpoises; there was the excitement of catching the occasional shark; flying fish were a welcome change on the menu. The Master, now that he had established his authority and his standards of seamanship, could afford to relax his grim visage; the once half-drunken rabble had become at once a community of individuals with marked characters and a highly skilled crew working almost instinctively together under the two Mates. Even more important than any harsh, formal discipline born of the Master's absolute authority, a close bond had grown between Harvey, his Mates and the hands, woven of trust and professional respect.

The doldrums, however, briefly interrupted this euphoria. For five days and five nights *Dalhousie* rolled horribly on the swell with no wind to lift her sails, to the nerve-wracking noise of a straining standing rigging, screaming pulley blocks, grinding braces and flapping sails. Rain fell continuously in a slow, enervating downpour. Harvey, anxious to catch the benefit of every faint zephyr to move his ship southward, was constantly calling orders to brace yards, and the men became sullen with fatigue from the heavy work in such high humidity, the constant handling of sodden ropes softening and splitting open the flesh of their hands.

These frustrations were quickly forgotten when the southeast trade wind emerged from a vicious rainstorm one morning and all was transformed to blue skies and a steady breeze again. Day after day, night after night the sea miles were reeled off. Then one day, in between Trinidad Isle and Cape Frio, Harvey realised that he was at the parting of the ways; it was here that he began to free his wind onto a direct course to take him west of the Falklands instead of pointing up, still on a taut bowline, on a familiar curve towards the Cape and Australia.

Next, the northeast monsoon of the southern summer began to make its influence felt, carrying *Dalhousie* well south of the latitude of

Rigs of Sailing Ships and Vessels

SAILS OF A SHIP RIGGED VESSEL

KEY

1. Flying jib
2. Outer jib
3. Inner jib
4. Fore topmast stay sail
5. Fore sail, or fore-course
6. Lower fore top-sail
7. Upper fore-top-sail
8. Lower fore-top-gallant sail
9. Upper fore-top-gallant sail
10. Fore royal
11. Main-sail, or main course
12. Lower main top-sail
13. Upper main top-sail
14. Lower main-top-gallant sail
15. Upper-main-top-gallant sail
16. Main royal
17. Main skysail
18. Cross-jack (brailed up)
19. Lower mizen top-sail
20. Upper mizen top-sail
21. Mizen top-gallant sail
22. Mizen royal
23. Spanker, or Driver
24. Main top-mast stay-sail
25. Main top-gallant stay-sail
26. Main royal stay-sail
27. Mizen top-mast stay-sail
28. Mizen top-gallant stay-sail

THE SHIP IS A VESSEL WITH 3 MASTS AND SQUARE RIGGED ON ALL MASTS

the Rio de la Plata on a fair wind. Then, as the seas grew colder and the winds began to puff up from the west, Harvey began to prepare for the testing times ahead: light weather sails were sent down and heavy weather sails, which had been worked over carefully during the passage of the trades, were bent on. The Mate checked the trim and stability of the cargo; hatches were covered with tarpaulin and battened down firmly to withstand heavy seas falling inboard.

South of about 35° south the environment was not unlike that of the North Atlantic, with westerly winds predominating but an almost continuous passage of depressions bringing about rapid and frequent changes of wind direction and strength. As *Dalhousie* approached the latitude of the Falkland Islands the frequency of gales increased, even though it was midsummer. Harvey, pacing the weather side of the poop deck for hour after hour, found himself continually asking the question whether she was carrying too much sail or too little; and at all times it grew rougher and wetter and colder, the sky more grey.

Concluding that the frequency of wind shifts to the south of west was such as to make the passage of the Straits of Le Maire imprudent, Harvey shaped his course from 50° South to pass well to the east of Staten Island on starboard tack. As *Dalhousie* cleared the latitude of Staten Island the full force of the mountainous Cape Horn seas struck her for the first time: the size and speed of the moving waters astonished even Harvey, who could only mutter to himself time and time again the words ‘Bloody Hell’.

The hands, clad in sou’westers, streaming oilskins and thigh-high seaboots, looked grave and tense as they eyed the sky, the rigging and, from time to time, the demeanour of the Master on the poop. The ship, close-hauled on starboard tack under topsails, clawed southward towards the ice. Harvey was familiar enough with the icebergs calved from the Arctic glaciers that drifted down into the Western Ocean; but here in the Southern Ocean the bergs consisted of immense islands, usually flat-topped, that had broken away from the great shelves of ice surrounding the Antarctic continent. Although at this time of year the pack-ice was well to the south, the risk of meeting the huge ice-islands was ever present. Lookouts were doubled.

Harvey now had to contend with a beat against adverse gales, the east-running Southern Ocean current and enormous seas for some 600 miles to the meridian 85° west before he could bear away northward. Apart from the ice hazard his main concern was that he should not find himself too far south when the wind blew from the north and northwest (which brought the severest gales at this time of the year—one day in four, as the succession of depressions drove eastwards from Cape Horn to South Georgia), nor too close to the land when it blew from south or southwest. The wind screamed in the rigging and the heavy seas crashed on board, and when the lulls came they were the treacherous precursors of sudden, unexpected shifts of wind direction and fresh hard blows.

Dalhousie battled well, if a little heavily, under her Master's skilfully judged handling. Some fifteen days after rounding Staten Island, sailed mostly under topsails and reefed foresail in frequent gales and colossal seas, she fetched longitude 83° west. The passage of Cape Horn could be said to be over. Harvey was fortunate to have so quick a passage with no problems from ice and no serious damage to his ship.

Thankfully he bore away to the NNW for Valparaiso, aiming to keep about 150 miles off the land, looking for the gentler southwesterly winds and the favourable current usually found in about 48° south.

Gradually the weather improved, bringing sunshine and a fair wind. *Dalhousie* nevertheless rolled fiercely in the long Pacific swell. After five days of steady, fast sailing on a broad reach Harvey reckoned that with the Humboldt current now under his counter it would be as well to close the coast to avoid being driven north of Valparaiso. When dawn broke the next day the lookout on the fore top picked up the Cordilleras 100 miles away, dominated by the 21,000 feet peak of Aconcagua. Shortly before noon on the day after, *Dalhousie* signalled her identity to the signal station on Curanmilla Point and received her instructions to anchor in Valparaiso Bay, a semicircle of water open to the north, sheltered except from the north winds that prevail in winter and bring in a heavy, rolling sea. So, on 20th February, after 93 days at sea, she came to rest a mile offshore with a thunder of chains as she anchored in the deep water.

Captain Harvey immediately handed the ship over to the Mate and, formally dressed in frock coat, white cravat and silk hat, had himself rowed ashore in the longboat to present the ship's papers at the Custom House and seek out his Company's Agent at the Exchange. The Agent took him to lodgings in the town which, rising like an amphitheatre round the bay, he found to look more English than Spanish in appearance. Of its population of some 50,000 a large proportion were European, and of these the English were the most numerous and commercially important. The part of the town known as the Port was built on the lower slopes of the hills surrounding the bay and on the narrow beach. Here, near the landing place, was the Custom House with its long, handsome facade surmounted by an elegant tower, and the Merchants' Exchange.

Many of Harvey's days were spent in the Reading Room where business was transacted and foreign intelligence was available.

To the west of the Port, where the base of the hills lay back from the sea, the former rural suburb of Almendral was fast turning into the principal part of the town, and here he lodged in one of the several excellent French hotels. For evening enjoyment he had the Theatre, one of the best on the continent, and the English Club. He found Valparaiso a congenial place where the Master Mariner was a highly esteemed personage.

His aim was to get away from Valparaiso with the best freight and while there was still a reasonable length of daylight in the latitude of the Horn, and well before the onset of the worst of the winter gales there, where the frequency rises to something like 20 days out of 30. His most serious concern was the increased risk of meeting icebergs in the South Atlantic; for this time, instead of beating slowly under reduced canvas against wind and sea, *Dalhousie* would be rushing helter-skelter downwind in the deteriorating visibility of late autumn snowstorms.

Meanwhile he had three tasks to perform during the couple of months which was the minimum he could expect to stay in Valparaiso: the first was to arrange the rapid offloading of the cargo he had just brought out; the second was to negotiate the largest cargo for the return passage; and the third was to explore the possibility of setting up an agency for his own firm of Watson and Harvey.

In the event, all was ready to sail for home in the last week of April, and *Dalhousie* cleared the signal station at Curanmilla Point on the 24th. The prevailing winds at that season being still from SE to S and the Humboldt Current setting northward, Harvey planned to make an offing as quickly as possible some 500 miles to the southwest, where he could be certain of picking up the westerlies and avoiding the adverse current. He headed close by Juan Fernandez Island to about 40° south, 85° west, standing thence on starboard tack with the west winds on the beam, on a southerly slant to latitude 50° south, whence he swung his ship gradually through SE to E so as to round the Horn in latitude 56° 30' south. Although the prevailing winds were favourable, they were also fierce. The current close under Cape Horn setting strongly to the east helped, particularly when he found it necessary to scud under bare poles. Though fast, it was a cold, wet, miserable passage under dark skies with frequent heavy snow and sleet showers, and Harvey was glad that he had sailed sufficiently early to miss the worst of the frequent easterly gales associated with the passage of the winter depressions. It was with relief that the Horn was passed and course set for a position 80 miles south of the Falkland Islands.

Thence *Dalhousie* sailed into the middle of the South Atlantic, riding first the westerlies to about 35° south, 35° west, then northward through the light variable winds of the Horse Latitudes and into the southeast trades in about 27° south. In the halcyon weather of the trade winds, the harshness of the Roaring Forties and the mountainous seas off Cape Horn were soon forgotten. The ship boomed along northward into the westerlies of the North Atlantic where once again the heavy weather canvas was set for the final fetch to Queenstown for orders.

She arrived there on 9th August, and somewhat to his annoyance Harvey was instructed to deliver his cargo in the Port of London. Having sailed on 13th, she bowled up Channel, lay waiting for the tide in the Downs on 18th, and moored in the Pool of London on 20th August. It was mid-September before she dropped anchor off the Tail o' the Bank in the Clyde, having sailed north about.

Margaret Harvey like most Captains' wives did not go out to meet her husband at the conclusion of his voyage, but waited until he came home. After nearly a year's living alone with her children, always dreading the arrival of a messenger with terrible news, her happy welcome for him was tinged with the knowledge that he had a difficult adjustment to make to being enclosed within four walls, while she—for her part—had to live with the constant dread that with barely any warning at all he could be ordered away.

Sure enough, on October 23rd another year's parting was upon them, and Harvey once again stood on the weather side of the poop deck as the tug pulled *Dalhousie* down river outbound for Valparaiso. He and Margaret parted with mixed feelings, sad that a mere four weeks was not long enough to get to know how to live together again. Harvey was somewhat relieved to escape from the land and the four walls to the reality of the lonely, masculine life of absolute command on board, the only place where he felt completely at one with himself. Margaret, too, took up the rhythm of her own life once again, keeping house at West Stewart Street and putting a brave face to the outside world despite the ever-present fear that at any time the sea might snatch her husband from her.

So this, until 1863, was the pattern of the Harvey's life: a month together at home in midwinter, followed by his year-long voyage to Valparaiso. They prospered, for he had an interest in his vessel, and as Master he was entitled to a proportion of the hold for his own ventures. Also contacts made in Valparaiso helped his own ship chandlery business to prosper further. He was a stout-hearted Mariner and a good man of business.

In May 1861 *Dalhousie* left the Clyde within a few days of a little wooden brig, *Anna Maria*, registered in Dumfries. The vessels spoke to each other once on passage, lay within sight of each other in Valparaiso Bay, and sailed for Britain at about the same time. The Mate of *Anna Maria* was a powerful, stockily built Prussian, blue-eyed and blond of hair, then twenty-seven years old. The son of a Private Secretary in Berlin, Gottlob Friedrich Wilhelm Linck inherited a family's wanderlust and began life at sea as a cadet in the Prussian Merchant

Marine. He became regularly engaged in the trade with general cargoes between the Baltic ports and Leith and with flax for Dundee to feed its dynamically growing business in linen and sailcloth. In Dundee he was made welcome in the house of a Jute merchant, James Wilson, and there in 1857 when he was twenty-three years old he first met their sixteen year old grand-daughter Margaret Dowie who was on holiday from Glasgow.

Friedrich promptly fell in love with her, quitted his Prussian snow and sought employment in a vessel working out of Glasgow so that he might improve his chances of seeing her between voyages.

The 1861 voyage in *Anna Maria* was Linck's first and only Cape Horn voyage, and a testing experience it was with gales on two days out of three as they beat resolutely to win every inch of westing to weather the Horn. The long hours of darkness, the spray that froze instantly on contact, the thick encrustation of ice on canvas and rope and block and tackle, the driving snow and sleet, the never-ending scream of the wind, the sickening rolling and the terrifying sight, hour after hour, of vast mountains of water, its surface whipped by the gale into the semblance of a seething cauldron, rising out of the gloom and bearing down remorselessly on the little brig were enough to bring men to the verge of breakdown.

That they persisted was due largely to the willpower of the Master. That they succeeded not only in surviving, but in reaching the declared goal, was due as much to the Mate's own seamanship skills and his ability to drive men beyond their endurance, as to the Master's judgement and decisiveness. Master and Mate were complementary, the one to command and the other to ensure that the work was done instantly and properly.

Although Harvey and Linck shared within a few miles and a few days of each other the physical and mental ordeals of that winter rounding of the Horn, they never met. Fifty years on, however, Harvey's grand-daughter Frances was to marry Linck's youngest son George.

When Captain Frank Harvey brought *Dalhousie* back to the Clyde again, it was evident that she was too small to be really effective in the

Chilean trade; so he agreed with Handasyde and Henderson that she should go into dock in the spring of 1863 to be cut in two and lengthened. The Company offered him command of one of their new iron screw steamships, the 300 ton *Dom Pedro*, in the Mediterranean trade, to try his hand at the new technology. He took up the offer with enthusiasm. On 14th August of that year he brought her back to Glasgow and by sheer coincidence his brother, Captain John Harvey, brought his ship from North America to anchor off Greenock.

The two of them spent Sunday together at Frank's house, 20 West Stewart Street, catching up on each other's life after a long separation, and on Monday morning parted in the street outside Greenock railway station intending to meet in Glasgow later in the day. Frank dashed up the steps to catch the 8.45am express to the City. When he reached the platform the train was moving slowly out. He tried to enter two compartments of a first class carriage but they were already full. He opened the door of the last compartment of the carriage and was running backwards along the platform in anticipation of jumping in when he collided with a pillar and was thrown onto the line between two coaches. The wheels of the carriage passed over him, killing him instantly. He was forty-nine years old.

The newspapers reported his age as being about sixty, a commentary on the physical effects of a harsh life at sea. Margaret's premature widowhood was probably no worse in the long run than the ordeal of old age shared with a worn-out ruined shell of a man, for it was to that that the tough life at sea reduced many of those mariners who did not die in harness.

12

The Coming of Steam

IN THE YEAR 1788, WHEN THE YOUNG SAILOR WILLIAM Watson married Isobelle Fleeming, there took place at Dalswinton, in Dumfriesshire, the first successful demonstration in the British Isles of a steam-driven boat.

The first people to employ steam for this purpose had been Fitch and Rumsey in America in the preceding year. In Scotland, under the patronage of Patrick Millar, William Symington now demonstrated an engine based on Newcomen's, but incorporating a condenser which effectively infringed James Watt's patent and to which he had applied his own patent method of obtaining rotary motion by means of chains and ratchet wheels. The Dalswinton trial was successful enough to warrant a second trial the following year with a bigger engine on the Forth-Clyde Canal and this, too, was successful, but Symington was forced to discontinue work for a number of years when Watt learned about the infringement of his patent.

After Watt's patent expired in 1800 Symington was able to adapt the best features of Watt's engine to his own unique way of converting the rectilinear motion of the steam-driven piston to the rotary motion of paddles, which eventually became the universal system for paddle-wheel steamers. He put the system into a little tugboat christened *Charlotte Dundas*, and in March 1802 she towed two barges for nearly twenty miles along the Forth-Clyde Canal against a stiff headwind. Although the trial was an outstanding success the Canal Company forced Symington to lay the vessel up because they were worried about erosion of the canal bank by the paddle wash. A disappointed man, he drifted into obscurity.

The makers of steam engines were struggling to keep pace with accelerating industrial demands, while the few people in shipping who had been inspired by the success of the *Charlotte Dundas* lacked the capital to follow it through. One Henry Bell of Glasgow and Helensburgh nevertheless quietly kept some design and development going, and in 1811 he was confident enough of the viability of his ideas to place an order for a small boat with John Wood and Son of Port Glasgow.

John Robertson of Glasgow built an engine for it to Bell's design; John Napier's of Glasgow made the boiler; and the young David Napier, later to become the most brilliant of the first generation of the Scottish Marine Engineers, installed the machinery in the hull. The vessel, named *Comet,* displacing about 25 tons, was launched early in 1812. Old deep-sea mariner William Watson, who watched her threshing the water off the Tail o' the Bank, thick black smoke pouring from the single high chimney that also did duty as a mast for auxiliary sails, looked on her as little more than a forty-foot open boat, possibly more effective though less reliable than a ship's jolly-boat for towing in and out of harbour. His son William, the young articled accountant, and his friends the Leckies, however, shared Henry Bell's enthusiasm and were not surprised when on July 6th she steamed up the Clyde to the Broomielaw to be received by a tremendous crowd of excited people.

The enterprising young folk nodded sagely when Bell put the *Comet* into regular ferry service between Glasgow, Greenock and Helensburgh. The journey between the Broomielaw and Greenock took about three hours, and she eventually achieved a scheduled service speed of 6 to 7 knots. Towards the end of 1814 even the elder heads nodded sagely when the paddle steamer *Marjory*, built in William Denny's yard at Dumbarton, was towed through the Forth-Clyde Canal and then, using her 30 foot funnel as a mast, sailed down the east coast to become the first steamboat to operate on the Thames. Two years later *Marjory* sailed to Newhaven and became the first steamboat to cross the Channel.

Bell's *Comet* steamed on the Glasgow-Greenock-Helensburgh service for four years. He then took her through the canal to work on the Forth for some time. In 1819 he established a service from Glasgow to Fort William until tragedy struck on 13th December 1820. On the way back from Fort William *Comet* was wrecked in a storm off Craignish Point in Argyll and became a total loss. As the accountant William Watson remarked, she may never have been profitable to operate, but at least she was the first steamboat to run a commercial service in Europe: and she sailed not at her Master's discretion, some time or other depending on wind and tide, but at an advertised time day in and day out. He reckoned that if his three sons William, George and Robert were to follow their grandfather's calling, he would like to see them as Masters in steam; and indeed, such was the astonishing pace of progress that all three were to become Masters of large ocean-going steamships on the North Atlantic.

Although in 1819 some Americans sponsored the first crossing of the Atlantic by an auxiliary steam-propelled vessel, the brig *Savannah*—using her machinery for an aggregate of about eighty hours on a 25-day crossing—it was mainly the British who drove the application of steam to the deepwater routes. The honour of being first to cross the Atlantic enirely under steam propulsion went in 1838 to the little Irish ferry steamer *Sirius*, chartered in a dramatic race to upstage Brunel's magnificent *Great Western* which he had designed specifically to capture from sail the British Government's contract for the mail packet service between England and the United States.

Despite Brunel's technical brilliance and early start, however, the mail contract was taken in 1840 by an alliance between Robert Napier and an entrepreneur new to Atlantic shipping, Samuel Cunard of Nova Scotia who, together with others, founded what was to become known as the Cunard Line. Their first four paddle steamers, all of them wooden, were built at Port Glasgow. Robert Napier himself became a shipbuilder at his Govan yard where he built a whole series of vessels for the Cunard Line, including their first screw-propelled ship *China* twenty years later.

In the 1850s steam was still playing a secondary role in long distance carriage of bulk cargoes and the 1860s saw the sailing vessel at its zenith with the legendary races of the China tea clippers. In the year before *Cutty Sark* was launched, British steam tonnage amounted to 170,000 compared with 4,500,000 tons of sail; but on the Western Ocean steam was already preponderant. The rediscovery of the dramatic fuel economy of the compound steam engine in the 60s hastened the decline of sail tonnage, but even so, parity between sail and steam was not reached for another two decades.

In 1866 Friedrich Linck, obtaining his first appointment as Mate in steam as much because of the connections between his Dowie in-laws and the Hendersons as to his previous excellence as Mate in sail, joined the 1700 ton iron screw steamship *Columbia*, pride of the fleet of Thomas and John Henderson of Glasgow. In the period between 5th December 1866 and 4th January 1868 *Columbia* made nine round trips between Glasgow and New York, each round trip taking about six weeks. This was an immense increase in productivity, but for the men it was wholly destructive of domestic life. Friedrich Linck's typical schedule in the steamship *Columbia* makes for an interesting comparison with Frank Harvey voyaging in his last two years in command of *Ellerslie* a decade earlier. Where Frank Harvey might spend weeks at his home port between voyages, Friedrich Linck spent no more than a day or two.

As the principal deck Officer Linck had no more than rudimentary knowledge of marine engines, nor did he need it, for *Columbia* carried all the masts, spars, canvas and hempen rigging of an orthodox sailing vessel, with additional sails fore-and-aft to steady her in a seaway when steaming. The simplicity of navigating crowded waterways like St George's Channel under steam in the teeth of a sou'wester appealed to him, but later when all plane sail was spread to a joyful wind he was appalled by the poor performance of the ship.

The old argument that it was necessary to be rigged for sail because of the unreliability of steam power had given way to the contention that the ability to sail reduced overall fuel consumption. Half a dozen voyages in *Columbia* convinced Linck that the coal saved whilst sailing inefficiently and slowly under canvas was more than offset by the additional coal needed to drive the weight and windage of the rig against the gales of the Western Ocean at all other times. Moreover,

The United Kingdom *(Edinburgh)*
steam vessel, 1000 tons, 200 horsepower

maintenance of the published, regular schedule on which the good opinion of the liner's clients depended did not leave much scope for dawdling under sail.

Whilst Harvey in *Ellerslie* and Johnston in *Pomona* took the Viking route via Cape Farewell to Belle Isle Strait and the St Lawrence, or the southern route dipping down to the Tropic of Cancer for the West Indies or the Gulf of Mexico, *Columbia* and her kind were able to take the direct great circle route to New York, right across the Grand Banks to the south of Newfoundland. Here the warm waters of the North American Drift curling northeast under a smother of mist meet the ice-laden waters of the Labrador Current sweeping down form the Arctic and, mixing, create perpetual fog. The Banks are one of the world's most prolific fishing grounds. The Master of a sailing vessel, under no pressure of a tight sailing schedule, navigated the Banks much as did the fishing skippers, particularly because fog was usually accompanied by light winds, and used his ears as much. In *Columbia* on the other hand, the sound of the engine and screws denied Linck the use of his ears while his Master, independent of the wind, maintained even in fog the high speed necessary to reach port on schedule. Time and time again the running down of a fishing vessel was only narrowly averted —and averted not by skill and judgement but by good luck pure and simple.

Like many a well-qualified Officer Linck rejected as unseamanlike the conditions of service in these Western Ocean Liners. At the earliest opportunity, being offered command of a small screw steamer, the *Grange*, trading between Glasgow and Spain, he quit the Western Ocean trade in the autumn of 1868.

His new employment brought him not only the fulfilment of his own command but also a week or two at home between voyages. Relatively, his family life flourished until November of the following year when, assuming command of a larger vessel, s.s. *Fitzpatrick* (942 tons gross) owned by Robert Donaldson of Glasgow, he once again came up against the rigidity of steamship scheduling which was a result of the combination of mechanical reliability and the cable telegraph system. *Fitzpatrick* traded from British ports to Spain and the Baltic. She cleared Glasgow for Bordeaux on 27th December 1869; she was in Cardiff for a couple of days in January; Glasgow for a week in April; Dublin for a couple of days in May; Dundee in July and two whole months in Glasgow at the end of 1870.

Shortly after Linck sailed to Huelva at the end of January 1871, his British naturalisation came through, and though he continued until the end of May to trade to Spanish ports he made up his mind to make a career for himself ashore so that he could settle and enjoy his growing family. His wife Margaret was a Dowie: her father, a Merchant Burgess, was a scion of a Glasgow family that ran a rope-making business and her mother Matilda came from a Dundee family of jute and sailcloth merchants. As a result of discussions with the Dowie in-laws, Friedrich Linck started his own ropery business in Spittal, but it was not a success. The family returned to Glasgow prior to 1877 when their youngest child George Emil was born—the lad who was eventually to marry Captain Frank Harvey's granddaughter. In 1879 Friedrich Linck died and was buried in Craigton Cemetery.

By this time all three of James Watson's brothers, William, George and Robert were Master Mariners sailing out of Liverpool. All three had been excited from an early age by the marvel of steam propulsion and William, the eldest, had seen his opportunity for rapid advancement in the steam packet trade that burst out of Liverpool in the 1840s. George was persuaded to follow him and serve as his Mate in *Lotus* in 1851 and 1852 before himself commanding her from '52 to '56. Robert, the youngest, was only twenty-four when he obtained his Master's certificate whilst serving as Mate in Liverpool steamers in the Mediterranean trade. Apart from his short spell as Master of the sailing vessel Waterwitch in 1853, his first command was the steamer *Orontes* (1856-61) followed by *Acadia* and *Amalia*, all three in the Mediterranean trade.

Liverpool was a port of no significance until it exploded into growth on the basis of the triangular trade with West Africa and the American Colonies: the carriage of cheap 'trade goods' to the Guinea Coast, of slaves on the 'Middle Passage' across the Western Ocean, and of sugar and molasses for the return to Liverpool. As the slave trade decreased under the pressure of the Abolitionist movements in the early 19th century, Liverpool—by then a highly developed port—was in a position to dominate the cotton trade with America and to hold a very firm grip on the African and West Indian trades.

The West Indies trade was more important in the 18th century than that with the East Indies in terms both of capital investment and of the quantity and value of commodities shipped. In 1790, for example, the respective investments were £70 million and £18 million. The West Indian trade with England was nearly double that of the East India Company; and in the ensuing 16 years over 650 ships sailed outward from British ports to the West Indies compared with only 63 for the East. The importance of this trade, however, diminished during the first half of the 19th century as a result of the great flow westwards of migrant people and eastwards of American cotton.

At the beginning of the American Civil War, cotton was Britain's most important single industry. Out of 13 million pounds of cotton spun in Lancashire, no less than 11 million pounds were imported from America through Liverpool. Its position opposite Ireland, moreover, brought it into even more rapid growth when, during the decades of vast migration to the New World after 1840, it came to dominate the passenger traffic on the Western Ocean.

After the outbreak of war between the American States disastrously cut off the huge inflow of cotton at source, the need to repair the damage to Lancashire's cotton industry led to an increase in supplies from India round the Cape and eventually, in 1869, to the opening of the Suez Canal. To all of this the ship-owners of Liverpool applied themselves industriously, but some continued to look towards the strengthening of the traditional trading links with the West Indian Colonies.

Captain Frank Harvey's enthusiasm for the potential of the Valparaiso trade which he followed for the last half dozen or so years of his life, coupled with the possibilities of opening up new trade with the Pacific Ocean by way of the land-link across the Panama Isthmus, no doubt had some effect on his young brother-in-law Robert Watson's decision to join the West India and Pacific Steam Navigation Company when it was formed at Liverpool in 1863. He remained with the company, sailing between Liverpool and the ports of the Caribbean Sea and the Gulf of Mexico, never actually entering the Pacific, for the next twenty-seven years until his death at Cartagena after commanding their steamships *Venezuelan*, *Barbadian*, *Cuban*, *Bolivar*, *Talisman*, *Californian* and others. All were subsidised to carry the Royal Mail.

A typical voyage out of Liverpool took departure from Cape Clear directly for Barbados 3,500 miles away, and having called at Bridgetown proceeded along the Spanish Main to Colon, calling at La Guaira and Curacao. The return passage was made by way of Cartagena, Kingston (Jamaica), Port-au-Prince (Haiti), the Windward Passage, the Caicos and Turks Island Passage and thence direct to a landfall at Cape Clear for Liverpool. The round trip covered about

10,000 miles. Another service included Havana, New Orleans and the Mexican ports.

Captain Watson was particularly well known in New Orleans, and after his death the Chaplain of the Seaman's Bethel there said of him:

> 'Captain Robert Watson is dead. The tidings have thrown a pall of gloom over our hearts. His was an active and a busy life . . . a long, useful and honourable life. We weep that we shall not again grasp his hand and enjoy his warm salutations. He was a Prince among seamen, a bold and dashing Mariner and an honour to his profession. When his vessel was announced the gladsome remark was made, 'Captain Watson is coming!' His visits were refreshing to people wherever he went. When I crossed the Atlantic in 1883 I well remember how people visited his vessel from the coast of Ireland, and as they came in their boats they exclaimed 'Watson! Watson!' He was a sea King and had his admiring subjects along all his extensive voyages.
>
> 'His life was strewn with generous, graceful deeds. Wherever he sailed, Captain Watson . . . stood by Bethels and Seamen's Institutes wherever he found them. He was a true friend of the New Orleans Bethel and its Chaplain. He attended it and was the life of those meetings when he did . . . I made his acquaintance in the epidemic of 1878. We were loving friends ever since. He took me to his house in Liverpool, introduced me to his family and friends and treated me like a brother.
>
> 'As a Public Carrier and Commander on the deep blue he had abundant opportunity of making people happy, and he improved it well. His passengers ever praised their captain: he was such a courteous gentleman to his lady passengers; the children all loved him; he was very much of a boy himself when off duty.'

The New Orleans Agent of the West India and Pacific Steam Navigation Company wrote:

> 'The death of our old friend Captain Robert Watson was an

exceedingly sad event to myself and an army of friends here who had the greatest esteem for the unselfish, kindly and true-hearted man that he was. Thirty years ago he was in command of steamers trading from Liverpool to the Mediterranean, but he joined our company when first formed twenty seven years since, and has been in constant command of the finest steamers of the line since then. His record is almost unparalleled, as in all this long service and navigation amongst the most dangerous parts of the West Indies he has never caused any loss to the company. This wonderful record, and the universal testimony of those Captains who have served under his command as Officers, prove him to have been one of the finest seamen who ever trod the deck of a vessel. His heart was in his work and all his energies were given for the company he served, whilst at the same time his love of justice was so real and his kindness of heart so great that no-one under his command ever suffered wrongfully, but oft-times wrong-doers received the benefit of his kindly heart. During an acquaintance with him of seventeen years and several voyages as a passenger on his vessel, I have been able to judge well of his character; and though as with all men he was not perfect, my judgement tells me that the world has lost a gentle-hearted, clear-headed man, and hundreds have lost a true-hearted, unselfish friend of whom their only remembrance will be one of sincere, deep sorrow.'

PART FOUR

East of Suez

13

Sailing Round The Cape

IN APRIL 1862, FRIEDRICH LINCK WAS STILL MATE OF THE *Anna Maria.* In that month she discharged her Chilean cargo at Liverpool and subsequently cleared outbound for Ceylon in June. She passed the meridian of the Cape of Good Hope well south of the west-running Agulhas current in mid-August, establishing her position by means of soundings on the Agulhas Bank. With a high probability of light variable winds in the Mozambique Channel at that time of year, the Master elected to stand eastward until 45° east longitude; then to swing to the northeast in a wide circle through the variables until he could pick up the southeast tradewind to carry the brig between Mauritius and Reunion.

Thence he steered northward between the Saya de Malha Bank and the Seychelles to the vicinity of the equator to catch the southwest monsoon, which gave him a free wind for Ceylon by way of the One and a Half Degree Channel at the southern end of the Maldive chain of islands.

Anna Maria arrived at Point de Galle in the middle of September. After discharging she lay there until the northeast monsoon, established in November, brought the first of the annual fleet of trading vessels from the China Seas. Having loaded a miscellaneous cargo of Far Eastern origin, the little brig cleared for home in January.

Course was set to westward to pass through the Nine Degree Channel at the northern end of the Maldives, thence to the west of the Seychelles and Amirante Islands. Still on a fair wind, *Anna Maria* sailed on between Madagascar and the Cormores Isles into the mainstream of the Mozambique Current and the Agulhas Current. For a brief period in the southern part of the Mozambique Channel she had to beat against southerly winds, but always the current gave her strong help on her way. It was early February, the best time of the year for making the westbound passage of the Cape of Good Hope, so once she had cleared the Mozambique Channel she was able to carry favourable easterlies until clear of the Agulhas Bank.

After weathering the Cape easily enough she soon pointed direct for St Helena, then Ascension. On meeting the northeast trades in about 5° north and 25° west, she sailed close-hauled until she lost them in 27° north. Negotiating the variables with reasonable despatch, she picked up the westerlies almost within sight of Flores in the Azores, where the colour of the sea changed from tropic blue to deep green, and was able to shape for the chops of the Channel. She reached London by way of the Downs on April 1st.

When he had seen to the unloading of *Anna Maria,* as was his duty, Friedrich Linck took his discharge and returned to Glasgow. His beloved Margaret Dowie, now twenty-two, was old enough for marriage. As soon as he had passed his examinations for a Board of Trade Chief Mate's Certificate he proposed to her and was accepted. He was then offered a berth as Mate of the iron full rigged ship *Severn* loading in Glasgow. The wedding took place on June 25th 1863, but unfortunately for Friedrich and Margaret the ship's owners decided only a short time beforehand to despatch the *Severn* to Plymouth to complete loading there. She sailed from Glasgow on the day after the wedding. From the moment when the Boarding Master bundled the new crew aboard stupid with drink and half destitute, Linck assumed the Mate's role of iron-fisted driver of men, for it was he who under the Master, Captain King, had to dominate them within minutes, to sift out the real sailors from the scum and engage their subconscious pride and skill.

Severn arrived at Plymouth on July 7th, where she lay for a week completing her loading. Margaret had travelled down overland but saw too little of her new husband for, busy holding off the crimps' runners who were attempting to use their hard liquor to wheedle the hands ashore to sell again as a crew to another ship, he had to supervise the loading and deal with the motley crowd of stevedores, lightermen, touts and all the dockies of one sort or another who made a living on the waterfront.

All persons coming on board must of necessity report to the Mate first. Out of the constant and chaotic coming and going amid the ever present personal abuse and threat of violence he had somehow to keep the conduct of the ship's business on board moving forward. He had a responsibility which made him potentially vulnerable to innumerable temptations of profitable racketeering. In foreign ports especially, the Mate's great danger in these circumstances was drink. Linck in common with the best of them forswore it, and this effectively cut him off from society ashore.

On 14th July *Severn* was ready for sea, bound for Australia. Linck bade his bride goodbye. The two did not meet again for nearly three years. Once in the northeast tradewinds, Linck saw to the sending down of the heavy weather canvas and the bending of the light sails, and daily supervised the overhauling of the standing and running rigging, the greasing of the masts and blocks, and the endless painting of the ship that was the routine in the tropics.

As Mate of a full-rigged ship he was at the very pinnacle of his skills as a seaman officer, and his first concern was for the care of the means of propulsion, the towering, complex structure of canvas, cordage and spars that needed unceasing attention. He developed a hawk's eye to see even from the deck the first hint of possible trouble aloft, be it only a loose end, a frayed seizing or a chafe. At sea it was his job to ensure that every man was fully occupied throughout the day's working hours: there was always work to be done; cleanliness, order and regularity were indispensable, and it was Linck's nature to have his ship as smart as any Man o' War with not a rope-yarn out of place, not a streak of dirt on her topsides, no unsightly stains on her masts, yards or canvas, and decks gleaming.

During the halcyon days in the tradewinds he saw to it that *Severn* was well cared for as though she were his hobby. Clad only in blue shirt, trousers, short boots and an old cap he was everywhere, working, overseeing, teaching, riding his hobby with an unflagging energy that infected all hands. His was professional leadership of a high order, for he was a perfectionist. The hands responded. Things had come a long way since, on the first day out of port, he had had to drive them with boot and fist and forcefulness of character.

The Mate was also responsible for keeping the Dead Reckoning independently of the Master. The two stood side by side taking the sun's altitude morning and noon, each independently worked his sights to a noon position. Linck handed his results to the Master who never divulged his working, then entered the agreed noon position in the log.

Having passed through the tradewind belt, course was shaped through the variables to pick up the wild westerlies as soon as possible, crossing the meridian of Greenwich in 40° south; whence, with yards squared and great grey seas crashing past from astern, *Severn* ran down her easting in that dark, stormy and cold latitude.

In that winter the wind, though seldom too strong for carrying sail, was enough to drive her in what could only be called a mad career. The main worry was the icebergs which are most numerous between New Amsterdam and Cape Leeuwin. Bergs of enormous dimensions: individuals up to 100 miles long and 1,000 feet high are not uncommon, and they can be encountered not only as low as 30° south but also in great numbers. So, threshing through the thick weather at high speed, every sail set that could be carried safely, two men were constantly posted on the fo'castle head, one pumping away on the foghorn and listening for the possible echo of an iceberg, the other on the lookout for small ice.

In mid October they sighted Cape Otway and shaped through Bass Strait north of King Island, sailed along the coast of Victoria to Gabo Island and then stood off to the eastward in about 54° east so as to avoid the full southerly set of the current along the coast of New South Wales. They made their final approach to Port Jackson from a point slightly to the north of it. In the last days of the month *Severn* entered between the heads on a light easterly breeze and ran down the beautiful sound of Port Jackson to the anchorage in Darling Harbour. A couple of days later she was brought alongside the circular quay at the head of Sydney Cove in the heart of the fair city.

The principal exports of Sydney in those days were wool, tallow, hides and the produce of the southern whale fishery; and because there was virtually no manufacturing industry in the State there was a very extensive import trade including grain, coffee, tea, sugar, rice, wine and spirits, soap and candles, wearing apparel, furniture, carriages and the whole range of manufactured goods such as hardware and agricultural implements. The discovery of gold in the colony in the early 50s gave an immense impetus to trade.

Here at Sydney *Severn* lay for a couple of months while her Master and many of her crew enjoyed all the pleasures of that fair city in its lovely setting, until towards the end of February she sailed for Auckland, still the capital of the young colony of New Zealand. A twelve day passage through the variables took her round the North Cape and down through the Hauraki Gulf to the deep-water anchorage off the city.

Large deposits of gold had been discovered on North Island two years before. The consequential large influx of people which brought prosperity to the pastoral colony nevertheless turned North Island into the focus of the Maori wars which intermittently flared throughout the decade 1861-71. On her two voyages between Auckland and Sydney in the autumn of '64 *Severn* was carrying stores in support of the Military garrison.

On 26th March, on her first passage back to Sydney, she encountered a cyclone a couple of days out from North Cape. While she was running under bare poles a huge sea carried away her starboard quarter rails and much of her port rails.

On May 24th, having unloaded at Auckland for the second time, she set sail for Calcutta. As the southeast monsoon was well established in the Arafura Sea and the southwest monsoon was beginning to build up in the Indian Ocean, Captain King chose to take her through the Torres Strait, for, once clear of the variable winds between New Zealand and New Caledonia, she would have fair winds all the way. He took the route outside the Great Barrier Reef into the western quarter of the Coral Sea, making for the Bligh Entrance to Torres Strait south of the Gulf of Papua.

About a third of the way through the channel, where it curved sharply to starboard, King ran his ship gently but firmly onto a coral reef and there she stayed a full twelve hours while the Mate and his men laboured fiercely to get her off. It needs seamanship and gruelling work to get a ship off into the teeth of the wind and against the set of the current, but Linck and his men did it, and under Linck's eyes the carpenter padded the strained hull plates so that *Severn* was able to proceed safely through the Arafura Sea, broad-reaching along the northern edge of the Southern Equatorial Current.

When she met the meridian of 85° west, King bore away to northward into the southwest monsoon and directly for the coast of Ceylon. Having raised the high ground of Ceylon on the horizon, he bore away and sailed through cloudy, rainy weather to make land just south of the mouths of the Mahanadi River. From there he took a line of careful soundings in the shoaling, sandy waters south of the mouths of the Ganges, proceeding through thick drizzle to the Pilot Schooner's Station on the ten fathom line at Sandheads, in the Balasore roads.

The entrance to the Hooghly was one of the most difficult in all India. Quite apart from the thick weather of the monsoon, the coast itself is low and hard to identify even when seen, and the water is so gradually shoaling that a vessel might be in ten fathoms while still thirty miles from the river mouth and out of sight of land. A Mariner unsure of his position only knew that ahead of him lay sand banks which, guarding the river mouth, stretched for miles to seaward; but once he discovered them, the channels through the banks were sufficiently buoyed and the pilot service was excellent.

Severn waited for the flood tide at Sandheads, when the pilot came aboard. He conned her along the channel through the banks into the broad estuary of the Hooghly. The tide, whatever the direction and force of the wind, had sufficient strength to carry the vessel up river, but the pilotage was tortuous and difficult on account of the frequent shifting of the shoals, the swirling currents and the rapid tidal stream.

At the command of the Master on July 26th Linck, in charge of the fo'castle head, ordered the carpenter to let go both anchors at Diamond Harbour, the old established port below Calcutta. Here the East India

Company's ships used to discharge their outward, and take on board the greater part of their homeward cargoes, and here in the steaming heat *Severn* was beached on a falling tide for a superficial inspection of the damage caused by the stranding in Torres Strait.

On Linck's advice Captain King decided that she must be docked at Calcutta. She took a later tide up river between its low jungle-clad banks. The rapid stream of the rainy season contending with the high rise of the tide rendered pilotage uncertain and dangerous, so that a high degree of skill was needed on the crowded waterway even though the river was still nearly a mile wide above Garden Reach.

As the southeast monsoon began to die out towards the end of September, by which time *Severn* had been repaired in dock, she took on a cargo of coal destined for the steamer bunkering station at Colombo in Ceylon. Owing to the lack of good harbours off the West Coast of the African continent the steam route to the Far East had been developed at first along the separate sectors of the Mediterranean and the Indian Ocean linked by overland transport. The former, after 1837, was an extension to Alexandria of the earliest steamship services to Portugal and Gibraltar; and the latter was a gradual working eastwards from Suez via Aden, Bombay, Colombo and onwards.

The development of this steam traffic greatly enhanced the number of sailing ships out of the coal ports of Tyne, Wear and South Wales, but any tramping bulk carrier such as *Severn* could find useful freight in the output from the developing coal fields of Bengal. In the middle of the century, before the opening of the Suez Canal at the beginning of 1869, more than 200 sailing vessels were needed annually to support the P&O steamers on services between Suez and Madras, Calcutta, Penang, Singapore, Hong Kong and Australia.

On 26th September *Severn* cleared Customs and dropped down on the ebb tide past Garden Reach and Diamond Head into the muddy Bay of Bengal. She made Colombo twenty days out from Sandheads, sailing through a mixture of light, variable winds under fine skies alternating with fierce squalls bursting out of black clouds. It was hard and tedious work, for every slant of wind, as often unfavourable as favourable, had to be used to advantage.

At Colombo she lay at anchor in the roads, unloading into and loading from lighters. By November when she sailed for the Hooghly again the northeast monsoon was well established. She had to work for her passage, making easting close-hauled across the Bay of Bengal into smoother waters and more moderate winds close west of the Nicobars, in the hope of being able to make the entrance to the Hooghly in one long tack therefrom. When she berthed at Calcutta on 29th December, the bracing cool season was at its height.

After six very pleasant weeks by comparison with the earlier steam bath, Captain King cleared out with coal for Point de Galle on the south coast of Ceylon. The ship arrived back in Calcutta on 21st April just as the hot weather was setting in at the change of the monsoon. The rains set in early that year. Friedrich Linck, nearly two years separated from his wife and engaged to a ship which appeared to have no prospect of returning to the British Isles, became very depressed as week after week went by and *Severn* swung forlornly at her moorings off Fort William. He determined to find himself a homeward bound vessel. On 26th June he sailed as Mate of *Blencathra*, but it was February before she arrived in the London River.

Meanwhile *Severn* unexpectedly picked up a freight for home and sailed on August 1st, calling at St Helena for water and fresh provisions on 5th November. The rest of her story is told in a report from Liverpool dated 3rd January 1866:

> 'The *Severn*, [Captain] King, Calcutta to London, experienced a four day hurricane and cut away masts on 24th December and was abandoned in a sinking condition on 28th, crew saved. A telegram reports *Severn* abandoned in 51° north 18° west. Crew saved by *R.B. Grove* arrived at Liverpool.'

The homeward passage in *Blencathra* was to be Linck's final voyage in sail. Once round the Cape running for St Helena, he put the watches to scraping, chipping and painting: the starboard watch to the starboard side and the mizzen mast, the port watch to the port side and the fore and main masts. The 2,700 foot peak of St Helena duly rose on

Captain Friedrich Linck
(1834 – 1878)

the starboard bow silhouetted against the skyline. Then *Blencathra* ran for Ascension, and the tradewind weakened as the high green island passed some twenty miles to starboard.

Once through the doldrums again she heeled well over, close-hauled to the brisk northeast trade with her lee scuppers full of water and heavy sprays sweeping over her weather rails, her Master aiming to make as little westing as possible before turning north and east for the coast of England.

The brisk, wet sailing put a stop to Linck's painting programme on deck until she ran into prolonged calms in the Horse latitudes off the Azores. Suddenly the horizon, empty for weeks, was filled with homeward vessels from all parts of the world, becalmed, only to vanish from sight as the variable westerly winds picked up and sent them speeding in a dash for the chops of the Channel.

As soon as *Blencathra* had unloaded in the London River Linck took his discharge from her. He spent the next nine months at home in Glasgow with his wife and twin daughters Wilhelmina (Minnie) and Matilda (Tillie) and begot a third daughter, Marie, while studying for and passing his examinations as Master in sail and in steam, and also his Extra Master's Certificate. Thereafter all his seagoing time was in steamships.

James John Johnston completed a couple of years service as Mate in steam at the end of 1877. Regulations required him to obtain qualifying seagoing time as Master in sail before being eligible to qualify as Master in steam. He was given command of the Glasgow registered barque *Skiddaw* (700 tons), built ten years earlier in Quebec. She measured 154 feet from stem to stern, with a beam of 54 feet and depth of hold of 20 feet. It was with justifiable pride that he stood on the poop surveying the length of her deck and her tall spars as she lay in the Tyne, even though she was heavily begrimed with the dirt of loading for a voyage to the East Indies.

Johnston cleared *Skiddaw* at the Custom House at Newcastle on 27th February 1878 and dropped her down on the tide to Shields. She sailed on March 1st. He elected to take her outbound through the English Channel and reported passing Dungeness on 5th. Four months later she arrived at Batavia, having taken the usual route down the Atlantic into the Roaring Forties, running down her easting as far as the 75th meridian, and then slanting northeast across the tradewinds towards Sunda Strait between Java and Sumatra.

It was a slow passage. He failed to make sufficient easting towards Christmas Island because he misjudged the effects of the southeast monsoon and the west-going south equatorial current, at their strongest at this time of year, so he fetched up well to leeward of Sunda Strait. Not only was it a long, hard, tedious beat to recover ground against both wind and current but also, when he made Tanjong Gideh, the point at the entrance to the Strait, he found variable winds from the easterly sector, tending towards strong northeasters during the middle of the day. Combined with two or three knots of adverse current, headwinds forced Johnston to anchor frequently to avoid losing ground, and for this reason he kept close under the Java coast. After four months out of sight of land the contrast provided by the verdant mountains and islands, and the dazzling colours, and the heavy coastal traffic of strange vessels of all shapes and sizes and weird rigs, was magical.

After weathering Anjer, Johnston took a fair wind eastward to the beautiful, crowded Bay that formed the open yet secure roadstead of Batavia and anchored there for twenty-four hours on July 4th. He went ashore to confer with the company's Agent who passed him orders to proceed some 300 miles further along the coast to Samarang.

Wholly entranced by the magic of the scene he took his time to sail through the tranquil waters close under the lush coast fringed with coconut palms, behind which the highly cultivated paddy fields rose gently to the foot of the great mountain chain, its flanks covered by vast forests of high, vigorous trees. He anchored for twenty-four hours off the little port of Cheribon under the lofty volcano Tjermi, and on August 14th dropped anchor in three fathoms in the roads some two miles off the port of Samarang.

There under the direction of the Mate the crew unloaded the cargo into native barges while James John Johnston stayed ashore negotiating for the return cargo of sugar. He found the walled town airless and very hot, but healthy enough because each day's tides covered the marshes in the vicinity and kept them fresh. Behind the town the ground rose rapidly to a considerable height.

On 11th September the anchor was broken out and all was set for the long haul home. Once clear of Sunda Strait *Skiddaw* dropped the Dutch pilot and reached straight across the southeast tradewinds to a position 200 miles southwest of Durban. With the help of the Agulhas Current and easterly winds on the Agulhas Bank she passed south of Mossel Bay on 13th November and spoke her position to a passing steamer. Two months later she arrived in Falmouth for orders. After four days there, she sailed to Liverpool to unload on January 21st, and there James John was met by his wife with the daughter, Ida Florence, who had been born in the previous July while he was beating through light airs and calm waters between the romantic Cheribon and Samarang. They were all together for just a week before *Skiddaw* left for the first of two voyages to the United States, the last that James John made in sail.

In 1862 Mr Alfred Holt of Liverpool experimentally ran a version of the compound steam engine even more economical than John Elder's. In the very year that the apotheosis of the high speed clipper ship was emerging at Greenock in the shape of *Ariel* and *Sir Lancelot*, the next-door yard, Scott's, was building three new steamships with Holt's improved engine for his China trade, so economical that these vessels could carry enough coal to propel 3,000 tons of cargo at ten knots over 8,500 miles without a stop.

Whilst *Ariel* in her fastest passage ever from Foochow to London took ninety days, Holt's ships would cover the route regularly and without strain in less than sixty-five.

For some decades yet, sail put up a hard fight to remain competitive on the long sea passages, but the competition bore as heavily on the mariner as it did upon his ships. Even by the middle of the century the Master had lost a great deal of his autonomy.

Once answerable only to God and his own conscience, he was increasingly at the beck and call of managers and Boards of Directors as world-wide telegraphic communications improved. All the petty bureaucracy of management—water clerks, shore superintendents, accountants and market experts—began to close in upon him. That restless, roving, ruthless man of courage the Master Mariner, who, armed with nothing but high skill and a forceful personality, drove generations of lowly seamen to conquer the hostile wind and sea, had to learn to conform.

The salary of Captain Frank Harvey or Captain James Johnston as masters in sail in the early 60s might be £20 a month; but by 1879 James John Johnston was earning barely £16 per month in command of *Skiddaw*, no more than Richard Woodget when he took command of *Cutty Sark* in 1884 and made their names immortal in the annals of speed at sea.

When Johnston became Master of the steamer *Newnham* he rated £20 per month for the first two years and £25 thereafter. But in 1893 fierce competition reduced even this to an initial £18 a month on joining, followed by increments to a maximum of £25 after seven long years.

14

Through The Canal

THE OLD PACIFIC STEAM NAVIGATION COMPANY, THE forerunner of the West India and Pacific Steam Navigation Company which employed the Watson brothers, was the first to use the compound steam engine that ultimately became the single most powerful cause of the eclipse of the sailing vessel. Its fuel consumption was one third of that of the single expansion engine. The guiding genius in developing it for marine uses was John Elder of Glasgow, who had served his apprenticeship with David Napier and teamed up with Charles Randolph, Scotland's leading millwright in his day. In 1856 the Pacific Steam Navigation Company added two new ships to their fleet, the *Inca* and the *Valparaiso*, fitted with the first Randolph and Elder compound engines. P & O was next to follow, then Cunard, and after 1870 this form of propulsion was installed in nearly every steamship built for the Atlantic service.

Its impact was even more marked on the traditional long distance routes to the Far East. When the East India Company's trading monopolies were terminated—India in 1813 and China in 1833—a long period of intense commercial competition was initiated; and as people realised that speed was no less important than carrying capacity, major developments took place in the construction of merchant ships.

Although the East India Company became interested in steam as early as 1822, and Lieutenant James Henry Johnston navigated the Paddle-steamer *Enterprise* from the London River to Calcutta three years later (113 days on passage, of which an aggregate of sixty-two days was under steam), immense difficulties hindered early competition by steamers on these long routes. At first the cost of providing intermediate stations was prohibitive, even on the short route via the Mediterranean, broken as it was by portage across the Suez Isthmus. The opening of the Suez Canal both saved the British cotton industry from extinction and gave steam its final impetus towards supremacy over sail.

About a month before Friedrich Linck sailed into the London River in *Blencathra* in 1866, Captain James Johnston of North Shields was there in his barque *Agenoria* unloading a cargo of grain from Constantinople. On board as ship's boy was the Master's thirteen year old second son James John, shortly to begin his sea apprenticeship. With the ways of the bulk cargo tramp-shipping trade of the northeast deeply ingrained in his inherited subconsciousness, by the time he was twenty James John had become Mate of the 400 ton barque *Amalthea* of Newcastle, well-versed in the trade of carrying coal to the coaling stations of the Mediterranean and East of Suez, and bringing back grain or coffee or sugar or rice and the like. Yet even then, steam colliers were seriously challenging sail on this route.

In a sense, James John Johnston was a child of the radical shipping legislation of 1849-50 which repealed King Charles' Navigation Acts (designed to safeguard British bottoms). With the inhibitions of monopoly trading swept away, British ship-owners reacted full-bloodedly to world competition and led the British Maritime Industry to the level where, at its peak in 1910, it dominated the world's trade. After 1850 a whole new generation of Master Mariners was shaped and moulded by the demands of speed, regularity and reliability not only on the Western Ocean but also on the long passages to Australia, out by the Cape and home by the Horn, and to India and the Far East.

James John made his first voyage through the Suez Canal in 1873 aboard *Amalthea*. She left Cardiff laden with bunkering coal on 7th July, made her offing in the usual way to 10° west and, close-hauled in a breeze from WSW, made for a landfall off Cape Roca. The wind held fair round Cape San Vincente and Tarifa into the Mediterranean and then fell variable, but she had a reasonable passage keeping to the easterly current along the Barbary Coast and thence more or less directly to Port Said where she arrived one month out of Cardiff.

Barque in London docks

After being towed through the Suez Canal she made sail again with a fair wind down the Gulf. As the southwest monsoon was still established in the Arabian Sea, *Amalthea* was able to run free the length of the Red Sea. The wind backing gradually from NNW to SW helped her round Bab el Mandeb and along the Hadramaut Coast to Aden, which she reached in the third week of August. She discharged her coal, proceeded to Bombay in ballast, and lay there from the middle of September until the middle of October awaiting a cargo of grain and the establishment of the northwest monsoon to enable her to sail westward again.

Taking departure from Bombay in mid October, *Amalthea* laid a direct course for the Mozambique Channel, leaving the Seychelles Group and Amirantes Island to port. Finding the monsoon weakening, however, she headed towards Cape Delgardo to catch the best of the Mozambique and Agulhas currents; then, as the wind swung through east to south and back to east, she stood down the coast far enough to give her a fair fetch over the Agulhas Bank. An uneventful passage brought her to Dublin for unloading on 2nd February, 106 days out from Bombay.

This, an extension of the pattern of the last decade of his father's life, was establishing the pattern of the great tramping fleets of the northeast ports carrying coal outbound to the bunkering stations that had to be established wherever steamships operated, and bulk cargoes such as grain or cotton for the return to Britain.

James John's next berth was in the 500 ton barque *Eastern Queen*—coal from the Tyne to Bahia and cotton or coffee back to Liverpool. He took his discharge from *Eastern Queen* at Liverpool towards the end of April 1875, and almost immediately after his return to Shields found a berth as Mate in the schooner *Auld Reekie* bound with a load of pitch from Middlesborough to Genoa.

On his second voyage in her he was shipwrecked off Seaton Snook while clearing the Tees for Barcelona. Driven by a gale onto the North Gar Sand after dark, she sank under the rising tide. The crew lashed themselves to the mizzen rigging and were picked up after nine hours by the lifeboat. It was a narrow escape, and James John made two important decisions, the first of which was that it was high time to marry his childhood's sweetheart.

For many years, close neighbours to Captain and Mrs James Johnston in North Shields were Captain and Mrs John Fleming. The wives, Fanny and Jane, were particular friends. John Fleming was the son of that Joseph who lost his life in 1859 through mistaking the new Godrevy light for the Longships and driving onto the Hayle bar at night in a full nor'westerly gale.

At Shields the Johnston and Fleming children grew up and played and went to school together, and at an early age the Fleming daughter Isabella Lillie was marked to be sweetheart to James John. But James John followed his forebears to sea and it was Isabella's fate to wait alone for him until December 1875 when the two dynasties were united in marriage at Tynemouth Priory church.

James John's second decision was that there was likely to be more of a future for him in steamships. Indeed, as steam established itself, ship owners were able to cull from the sailing fleet a fine body of officers capable of fostering and expanding that splendid discipline of the Merchant Service which was a by-product of the great era of competition in sail.

Having acquired his certificate of competency as 'Mate in Steam' in November 1875, James John obtained a berth as Mate of the 1100 ton steamer *Harold* in the following March. She was an iron vessel registered in North Shields, with a 99 HP compound engine, and she looked as much like a sailing vessel as it is possible for a steamer to be, for she had three masts, an insignificant funnel, a clipper bow and a bowsprit. Indeed she crossed yards on her foremast and could be said to be barquentine-rigged. She was flush decked.

Her Master hated engineers simply because they were engineers. He was of the school that believed engines were a passing craze, and he used to hoist sails not only to steady the vessel in a seaway but because, when there was the slightest favourable wind, he was convinced that they added to *Harold's* speed, which at best was about eight knots. James John never noticed any difference in the speed but, being sufficiently interested to speak surreptitiously to the engineers when the Master was asleep in his cabin, was prepared to acknowledge that in certain circumstances under sail less coal was burned while maintaining best speed.

Unlike the older generation of Masters who regarded engineers as unspeakable and the stokers in the depths of the ship as something sub-human, James John, even though this was his first voyage in steam, was not unfamiliar with the clannish engineers. His father-in-law John Fleming was connected by marriage with the Tyneside Chisholms. Old Stephenson Chisholm had been one of the first young seamen to serve in steamboats on the Tyne, and his vast progeny had followed him either as mariners or as engineers in steamboats. When James John Johnston took his deck apprenticeship before the mast in sail, his childhood playmate Stephenson Chisholm the younger took his apprenticeship in an engineering works. Only when this was completed did Stephenson go to sea, simply to pursue the profession of engineer in the environment of a ship rather than a mine or a factory or the railway service. This was typical of the wide division betwen the mariner and the seagoing engineer which lasted until the virtual disappearance of sail in the next generation.

As soon as *Harold* had steamed over the Tyne bar laden with coals for London, her Old Man ordered all plain sail, and despite the eternal thumping of the engine and the screw James John found his first watch on the passage to Yarmouth roads not unlike any other he had made under sail, for there was plenty of seamanlike things to be done with ropes and canvas and the cleaning down of the ship. But the passage from the Shipwash, against a fast tide into the teeth of a brisk sou'wester with all sail furled, through the East Swin and the Narrow Deep, was a revelation to him. His only real anxiety was to avoid collision with the endless succession of sailing craft running outbound on fair wind and tide, manned and commanded, it seemed, by idiots who did not appreciate the problems of a watch-keeper on the bridge of a steamer.

After discharging in the London River *Harold* lay in the Downs for a couple of days while James John supervised the cleaning out of her holds, then proceeded to Calais to load a miscellaneous cargo of machinery and equipment required for works on the Suez Canal. As Mate, James John was directly in charge of loading and unloading and preparing the holds, and he was quickly beginning to learn that operating a steamship, with its relatively quick passages and rapid turn-round in port, was not simply a matter of standing idly on the bridge while the engineers and stokers did all the work. It was with relief that, before going below, he watched Calais mole drop astern and heard the Master, having cleared Cape Gris Nez, order course to be set for the Casquets into the teeth of the prevailing wind.

Twenty-four hours later, giving Ushant a wide berth in thick weather, course was laid directly for Cape Finisterre, making allowance for the general easterly set which made a landfall there dangerous except in clear weather. With a westerly wind, staysails and spanker were set to steady the vessel against the roll of the long Biscay swell. And so, to the accompaniment of the steady thump of the great pistons, the miles were eaten up relentlessly and the great Capes one by one swept past at a much closer range than would have been thought prudent in a sailing vessel—Vilano, Finisterre, Roca, St Vincent, Trafalgar, Tarifa—as the ship hugged the southern coastline in fine weather. The scent of the pine trees and the eucalyptus was borne off the rugged mountains on the gentle land breezes of the night watches.

At Tarifa, white farmsteads and old Moorish towers were dotted on the mountainsides; across the Straits the serrated mountains of the Riff were a splendid spectacle.

Once through the Strait *Harold* made use of the east-going current, taking a line ten miles north of the Isle of Alboran, then following the Algerian coast, on through the Galite Channel past Cape Bon, Pantellaria and Gozo until she could lay a course direct to Port Said. On the fifteenth day out, Port Said hove in sight on the low, narrow, treeless strip of land that bordered Lake Menzelah. A pilot came on board to take her in.

At first light the following morning, steaming at 'dead slow', *Harold* started on the eighty-five mile transit of the Canal. From time to time she had to lie by at a buoyed passing station to let traffic pass in the opposite direction, spending seven or eight hours waiting for oncoming vessels on top of the seventeen or so hours of steaming time on passage. As no movement was permitted on the canal during the hours of darkness, it was not until late afternoon of the second day that she passed into the Gulf of Suez and dropped anchor to await her turn to unload. Soon the coaling barges came alongside; the ship was submerged under a deafening tide of chattering Arabs sweating and encrusted with coal dust as they ran like ants backwards and forwards along the planks between the ship's coaling ports and the barges. Their Egyptian taskmasters thwacked them with canes as they went, to make them run faster, and yelled all the while at the top of their voices. As night fell, the scene illuminated by naphtha flares was like a vision of hell.

On 22nd April *Harold* sailed in ballast for the Black Sea. Having dropped the Canal Pilot at Port Said, she steamed directly for the entrance to the channel through the Dodecanese Islands, between Rhodes and Carpathos, thence between Mount Elias and Chios and close by Mytilene into the Dardanelles. Conning the ship through the Bosphorous into the Black Sea, James John pondered how his father and his father-in-law John Fleming had so often made this passage, and as an icy wind from the Ukraine forced him to hunch his shoulders behind the canvas dodgers he thanked his stars for the genius of James Watt and his successors who had made it possible for him to enjoy the relative comfort of a steamship. Six days out from Port Said they picked up the Danube pilot and steamed a short distance up-river to Killia where in due course *Harold* loaded a cargo of grain.

The return passage after leaving the Dardanelles took her through the narrow channel between the Cyclades and the Greek mainland, then through the Cervi Channel north of Kythera and across to a landfall at Cape Passero, the southern tip of Sicily. *Harold* reached Falmouth on 25th May and discharged the grain at Cardiff on 30th. That year she made three further similar voyages, after which James John decided to accept an offer to take command of a barque in order to acquire sea-time as Master in sail before qualifying to command in steam.

In 1877 Hatfield, Cameron and Co. gave him successive commands of the Glasgow registered barques *Corsica* and *Skiddaw*. As Master of *Corsica* (800 tons), he made one voyage to Buenos Aires and one to Savannah, Georgia. In *Skiddaw* he made the voyage to the Dutch East Indies already described in Chapter 13.

His next berths were as Mate in the screw steamers *Ben Nevis* in the Mediterranean, and *Laleham* for a couple of voyages to U.S.A., after which he was given command of s.s. *Newnham* (2,000 tons) for a single voyage. In 1881 he voyaged to the East Indies again, this time prosaically in a steamship. He made two such voyages, each lasting about three months, the first as Mate of s.s. *Argosy* and then as her Master.

This was followed by a voyage to North America before he took *Argosy* all the way to Samarang for the third time. At Marseilles on the way back early in the following year he was transferred to s.s *Newnham* in command, an appointment which he held for the next six years. She was an iron screw steamship of 2,000 tons. From Marseilles he took her out East and then to North America, and back to London in August. This broadly was the continuing pattern of her trading, but in the spring of 1884 she set out on a voyage to the South Pacific at the opposite side of the world.

What distinguished James John Johnston of North Shields from Frank Harvey and Friedrich Linck on the Clyde, and Robert Watson on the Mersey, all of whom were operating fixed schedules on regular routes, was that as Master of a bulk tramp ship James John had the whole of the Seven Seas for his parish, taking the Indian Ocean and the South Pacific in his stride with as much ease as the Western Ocean and the South Atlantic.

A typical spell followed his appointment to command the *Newnham* in February 1882. He took her to the East Indies, thence to the Atlantic seaboard of the USA and on to London. He then made two round trips between England and the USA. In February 1883 he steamed to the East Indies by way of Suez. After three round trips between Cardiff and the Indian Ocean during the year, he steamed through Suez, round the Cape, up to the Atlantic coast of USA and back to Liverpool in the spring of 1844. It was, of course, not coal alone that *Newnham* carried outward; larger and larger quantities of ironware, rails, locomotives and trucks, and machinery of all kinds were being sent abroad, finished manufactures in exchange for raw materials.

On 5th April 1884 *Newnham* steamed out of Cardiff with a cargo of coals, transited the Canal about the 21st and reached Colombo on 8th May. It being in the inter-monsoon period, the weather was very fine all the way from Suez except for a few isolated squalls towards the end of the passage. James John took the direct route from Aden between Cape Guardafui and Socotra, and on through the Eight Degree Channel which separates the Laccadive Islands from the Maldives. By the time *Newnham* left Colombo in ballast on 14th May the southwest monsoon appeared to have set in, so that it was a rough, wet passage round Dondra Head; but once around, with wind and sea astern, she had an uneventful run into the Bay of Bengal, and after picking up the pilot off

Sandheads arrived in Calcutta on 20th May.

A month later, laden with coals, she sailed for Fiji. With the monsoon now well established, James John decided to take the sheltered route east of the Andaman and the Nicobar Islands into the Malacca Strait. There was a considerable amount of traffic in this narrow passage, including several large fishing fleets. The visibility was, for the most part, poor with a great deal of heavy rain, so the Master spent most of his time on the bridge during the three-day passage of the narrowest part of the strait approaching Singapore.

Beyond Singapore, steaming through the Karimata Strait into the Java and the Flores Seas, *Newnham* encountered the first of the southeast trades which freshened gradually as the voyage progressed. Visibility remained poor all the way into the Arafura Sea. James John conned her through the Torres Strait and out by the Great Northeast Channel into the Coral Sea.

Day after day the trade wind blew hard, though it let up a lot at nights, and *Newnham* pitched deep into the long head seas as she steamed along the line of the Louisiade Archipelago. On the third day after leaving the Louisiades, having passed beyond sight to the southeast of the Solomons, the Mate on watch picked up the high peak of Vanikoro Island at the southern end of the Santa Cruz Islands and, shortly afterwards, the lower peaks of the Torres and Banks groups at the northern end of the New Hebrides. After passing between Vanikoro and Torres, a course was laid south-eastward towards Viti Levu, the largest, highest and principal island of the Fiji group. *Newnham* negotiated the Kandavu Channel and arrived at Suva on 21st July. James John considered this passage from Singapore to Suva through some 3500 miles of volcanic islands and coral formations to be one of the most interesting and demanding pieces of sustained pilotage in his whole career at sea.

Suva, on the eastern side of a superb natural harbour which was protected by coral reefs stretching almost fully across its entrance at the south and by high ground from the prevailing winds, had been the capital for barely two years when James John visited it. In annexing the Fiji Islands in 1874 the British Government was primarily concerned with the establishment of a coaling station between Panama and Australia.

Prior to that the luxuriant islands, mountainous and possessed of a wonderful climate, had been notorious for the cannibalism of their inhabitants. Not only shipwrecked people were eaten up to the 1850s, but also friends and, indeed, dearest relatives as much as enemies—not for any religious reason, but simply because the Fijians relished the taste of 'long pig'.

The islanders excelled their neighbours in the useful arts, especially in building large canoes sometimes more than 100 feet long, and the construction of houses from natural vegetation, as well as the manufacture of native cloth and pottery.

From Suva *Newnham* followed a great circle direct to Newcastle, New South Wales, passing within a score of miles of each of three unlit hazards, Conway Reef, Hunter Island and Elizabeth Reef. She lay at Newcastle from 13th to 30th August to load coal for Singapore. Despite the odds on heading into westerly gales in the Australian Bight, James John decided to proceed south and west about Australia, passing within a score of miles of Christmas Island on the way to Sunda Strait. Singapore was reached twenty-two days out of Newcastle, and there the Chinese coolies, working in such perfect silence that one could hear the tread of their shoeless feet, unloaded the coal, carrying it in huge baskets each slung on a pole between two of them. Every tenth basket was weighed and an average struck.

On October 1st *Newnham* left Singapore to steam through the dying monsoon by way of Malacca Strait to Calcutta where she arrived a week later. She lay in the Hooghly River for a month before she was ready to set off in the clear weather of the northeast monsoon for Demerara by way of Table Bay (8/9th December) and Saint Helena (17th December). She tied up at one of the Water Street wharves at Georgetown a couple of days after Christmas and then cleared out on January 6th 1885 for New York. Northwest of Bermuda she encountered a severe gale that brewed up a tempestuous sea, but she arrived safely on 17th, unloaded and went to Baltimore in ballast. Dublin was her next port of call on 24th February, and after that,

London, whence she sailed again on 23rd March for the colourful East Indies. This pattern of voyaging continued for the next two and a half years.

During 1889 Dixon & Company of Middlesborough built a steamship of 2356 tons gross register (1526 net) for the Temperley Steam Shipping Co. Ltd. Completed in December of that year, she was named *Tunbridge*, and James John Johnston assumed command for her initial trial trip towards the end of that month. He then took her to Blyth, where she loaded coal for Alexandria. She sailed on 4th January 1890.

The weather across the Bay of Biscay was wild even for that time of year, and there was continual trouble with the engine. Notoriously the coast between Cape Ortega and Cape Finisterre is a dangerous landfall when strong onshore winds exacerbate the general easterly set in the bay and bring thick visibility and low cloud in the immediate offing. *Tunbridge* was set too close in as her steaming speed faltered.

Although the gale had abated, on the 10th January she stranded in thick weather off Cape Vilano. In the heavy swell she began to break up. The crew abandoned her, taking to the boats. All hands were saved by fine boat-work and were subsequently picked up by a passing vessel which landed them at Huelva, whence they were later shipped home as distressed British seamen.

Once safely ashore, despite the knowledge that his cool leadership was responsible for the survival of his crew, James John, on whom the stigma of a Master who had lost his ship by stranding would lie for the rest of his life, was overcome by a mental state of deep shock. The loss of *Tunbridge* broke his spirit, though outwardly for a few years he showed a bold and even dashing front to the world. He lacked the will to ask for another Master's appointment and refused to contemplate sailing in a subordinate position. He eventually sought solace in the bottle. He became unemployable and his marriage foundered. He died in his fiftieth year.

His seven children, despite all of this, were decently schooled and given a fair start in life. His last child, Ernest Livingstone Johnston, born in November 1891, excelled as a pupil at Tynemouth High School and in 1904, despite strong objections from his widowed Mother, was apprenticed to the Stag Line, a bulk carrier fleet which had been started by the Robinson family in North Shields around 1846. The Robinsons were one of those fine little family firms in the northeast who, by careful attention to tramp building and owning, ensured that the highest standards were brought to bear on the construction of bulk cargo carriers that were at once cheap, roomy, economical and seaworthy. They insisted that their Masters concerned themselves with training schemes for the apprentices, they nurtured sound officers and they treated them fairly. Their vessels were run by British officers and manned by lascars.

When Ernest Johnston sailed out of North Shields as the junior of two apprentices, the typical complement of a Stag Line steamer was Master, Chief Engineer, two Mates and two Engineers (these keeping watch and watch about), half a dozen lascar seamen under their serang and two tindals, and rather more firemen and trimmers. The two apprentices bunked and fed alone in a dog-hutch of a cabin of their own, and each was allotted to the watch of one of the Mates. The first duty of an apprentice was absolute obedience, for he had come to sea to learn and he could only learn by obeying.

Whilst all too often in those days, as in the preceding generations, the premium apprentices were treated as inferior cheap labour and driven harder than any others of a ship's company at the more uncongenial tasks of running a vessel, Ernest was fortunate to sail under a conscientious Master who felt himself in honour bound to look after the best interests of his apprentices.

For his part, Ernest was from the very first smart and eager to learn. Capable of his fair share of devilment ashore, he had a natural habit of ready and cheerful obedience afloat, an alert and well-tutored mind and the unashamed initiative to inquire what he might usefully do. On his initial voyage from the Tyne to Bombay he learned to know his ship thoroughly, on deck and aloft, by day and by night; to take his trick at the wheel and to help in the setting and the furling of sails—for even then the few sails that a tramp steamer carried were set whenever the wind was favourable or the Master imagined that they could be of the slightest advantage to the economy of steaming.

From the serang of lascars he quickly learned the lingua franca of the native seamen—'lascari bhat'—and acquired a fine fluency to curse in it. As junior apprentice he rated lower than a lascar ordinary seaman and must take his tutelage in the basic skills of a seaman from the tindals—something of a change from being a favoured pupil at Tynemouth High School. Nevertheless he was a white boy and a future officer who must maintain his racial superiority. But it needed more than the ability to curse in lascari bhat, more even than a proficiency with his fists, for a lad fresh from High School to establish a commanding position in a crew of jabbering lascars.

In a tramp ship east of Suez there was no Master-at-Arms backed by a company of Marines to enforce discipline, but only professional knowledge and skill, in which Ernest was at first deficient, allied to strength of character and a powerful pair of fists, both of which he possessed in plenty.

His mentor the second Mate might conscientiously teach him the professional lore of a mariner, but he reckoned it was no job of his to help a boy stand on his feet and become a man.

Nevertheless, by the time he reached Bombay the well-read, cultivated schoolboy was well on the way to becoming a tough-minded driver of men.

With one short visit to Tynemouth in 1908, Ernest's sea education proceeded over the next four years in a series of voyages spanning the Indian Ocean, the Arabian Sea and the Bay of Bengal, taking in Mombassa, Aden, Karachi, Bombay, Calicut, Colombo, Calcutta, Rangoon, Moulmein, Penang and Singapore. First he had to master the ordinary seaman's skills with wire and hempen rope, the care and management of standing and running rigging, the bending and unbending and setting and furling of sails, and the elements of handling a steamship when under canvas.

Taking his spells as lookout-man initiated the long process of learning through experience the intricate practicalities of the Regulations for the Prevention of Collision at Sea in limitless combinations of circumstances. He had to learn how to manage ship's boats in heavy weather; the marking and use of lead lines and log lines; the use of rocket apparatus in the event of being stranded; the construction, use and action of sluices and of water ballast-tanks, winches, donkey-engines and other deck appliances, engine room telegraphs and patent logs.

He had to become adept not only at the morse and semaphore codes but also acquire a working knowledge of flag signals and the British Signal Manual.

Later on his learning broadened to include elements of ship construction and stability, some of the more arcane elements of the merchant Shipping Acts, the regulations and the practical skills for stowing all sorts of cargoes, and even the elements of marine engineering (though the deck Officers and the engineer Officers still messed separately, small as the ship was).

For his theoretical studies of pilotage in coastal waters and deep sea navigation, Ernest attended the navigation school in North Shields in between voyages in the winter of 1908, no difficult subjects for a bright lad to absorb even without the well-used sea-time that he had behind him. Here he formally consolidated the theoretical knowledge needed for passing the examination for a Second Mate's certificate of competency. On completing his indentures in December 1910 while his ship was lying in Bombay, he appeared before a number of grave and senior Master Mariners who severally examined his competence to hold such a certificate. He satisfied them at his first attempt, whereupon his own Master signed him on immediately as second Mate.

Ernest soon began to feel he was the hardest-worked man on board. When the ship was at sea he not only kept his watch on the bridge four hours on and four hours off, but was also the ship's navigator with special charge of compasses, charts and chronometer; and in port his place was in the holds seeing to the stowing and unloading of cargo and coping with native stevedores whose one aim in life was to do nothing, or failing that to do wrongly the least amount they could get away with.

The proper stowage of cargo was as essential for the safety of the ship as was the proper navigation. It was a filthy job, especially when coal was the cargo; nevertheless photographs of him on deck at this period invariably showed him as a somewhat dashing figure in immaculate white ducks.

Back in North Shields, in the early summer of 1912, Ernest obtained his First Mate's certificate and then shipped as Second Mate again for a further eighteen months east of Suez in order to acquire sufficient seagoing time to qualify. Early in 1914 he sailed out of the Tyne again, this time as First Mate of the Stag Line steamship *Stephanotis* (4,000 tons). By then, he held a Sub-Lieutenant's commission in the Royal Naval Reserve as well, for like most of his generation he perceived the shadows of an impending war looming closer.

As Mate, Ernest now had responsibility to the Master for the overall running of the vessel and the discipline of the crew while at sea, as well as standing watch and watch about. In port he also had responsibility for the entire conduct of the ship's business on board as distinct from that which the Master must needs perform on shore. His was the job of looking after the shipment of all cargo, examining bills of lading and tallying, while his second saw to the stowage below.

When the ship lay alongside a wharf in a foreign port his overworked life was full of trouble, with a lascar crew devoting all its cunning to doing as little work as possible, a motley gang of stevedores capable of working only under the strongest and most constant compulsion, and a horde of touts, crimps, runners and idlers of one sort and another to keep at bay. Like the generations of his forefathers who had served as Mates before him, his work must be done in an environment where personal abuse and violence on the part of drunken seamen bearing a grudge, or dock-rats caught stealing, was commonplace. He could retain mastery only by knowing when to apply diplomacy or when to throw a punch to quell trouble before it overwhelmed him.

The training he had as Mate of a tramp steamer in the Indian Ocean brought out all his sterling qualities—courage, patience, persistence, intelligence and the orderly application of thought, the ability to get obedience from men in a tough environment, enthusiasm for doing a job well, personal endurance of physical hardship, the enjoyment of comradeship and, in the quiet moonlit watches on the gentler seas, a persistent sense of the romantic.

While at Bombay in July 1914, Ernest obtained his Board of Trade Master's certificate in steam. During the year, *Stephanotis* carried five cargoes of coal from Calcutta to Bombay, three to Penang and one to Calicut; then in May 1915 she loaded 6475 tons of wheat at Karachi and sailed on the 12th for England. After bunkering at Aden on 18th, Port Said on 28th and Gibraltar on 8th June, she arrived in the Humber on 15th and completed discharging on 18th June. When he had seen his ship unloaded, Ernest took his discharge and sought active service with the Fleet. This was a decision which was to change completely the direction of his life and lead him to a high reputation and an early death.

Ambitious, enthusiastic, articulate and handsome as well as being a hard driver, he had already discovered in himself not only an outstanding aptitude for practical navigation but also what was much rarer, particularly in the deck officers of his day, a quick and intuitive grasp of the theoretical and mathematical basis of the art.

He was appointed Navigating Officer of the mine-sweeper H.M.S. *Spanker* in July 1915 and served in her until the following March, when he volunteered for 'hazardous duties' in the Royal Naval Air Service. As soon as he became a fully qualified airship pilot he married his childhood sweetheart, Janita Daisy Wood, towards the end of 1916. He served with distinction as an airship pilot in war and peace until 1922, when his short service commission in the Royal Air Force was terminated.

With a wife and two children to support he had, reluctantly, to turn to the sea again to earn their bread. Robinsons were glad to have him back as Mate of the Stag Line's latest acquisition, s.s. *Ixia*, at a princely salary of £21.10s a month, just about enough to pay the rent of the flat they had been living in in London.

Ernest made four voyages in *Ixia* in the following twelve months. She sailed from the Tyne on 4th August 1922 laden with 4,000 tons of coal for Piraeus, thence in ballast to Algiers to load nearly 5,000 tons of iron ore for Rotterdam, and on to Cardiff to load 4,000 tons of coal for

Aden on 4th October. After discharging the coal she steamed in ballast to Beira to load a cargo of dates, wheat and barley for London. On 2nd January she sailed out of London for Barry to load more coal for Piraeus, whence she steamed in ballast, passed through the canal on 6th February and arrived at the Seychelles on 19th. There she anchored for nearly a month inside the coral reef of Port Victoria on Mahe Island for a long, slow loading of 4,500 tons of guano and a couple of hundred tons of copra.

Back in home waters early in April she off-loaded half the guano and all the copra into lighters in the London River, and the rest at Birkenhead. Once again she loaded coal at Barry for Aden, and from there she steamed in ballast to Karachi to load 5,000 tons of wheat between 26th and 29th June. She reached Avonmouth on 5th August.

Ten days later, having seen his ship safely unloaded, Ernest took his final discharge from the Merchant Service to accept an appointment with Daimler Airways, one of the pioneering commercial air lines.

The frustrations which Ernest had felt as Chief Officer of a tramp steamer are clearly implied in the tone of the following letter which he wrote from *Ixia* in the winter of 1922 to his former colleague Frederick Tymms, with whom he had worked for two years in the Air Ministry's Contollerate of Civil Aviation.

That had been an intellectually challenging life infinitely removed in style and content from the endless monotony, the sheer hard grind and often brutality of *Ixia*.

'So sorry I didn't get a chance to answer your very welcome letter from Port Said, but as a matter of fact we were only there for a few hours and I was terribly busy looking after the discharging and bunkering, and of course like all ships we put to sea at night after a hard day of it.

'The navigation tables you are asking about are Goodwin's Alpha, Beta and Gamma Tables published by J.D. Potter, price 8 shillings. My opinion of them remains the same and I've used nothing else since I came back to sea and have always obtained the best results with them. A further recommendation for these tables is that the 'Old Man', who is one of these rule of thumb fanatics who works all his sights to seconds of arc and steers course to half a degree, has no use for them and won't even listen to me explain them to him, but nevertheless always has me take sights with him morning noon and night, whether it is my watch below or on deck, so as to check his. It is really remarkable though that the old blighter flatters himself as being a smart navigator—I'll admit that he is above the average tramp skipper—but for all that he works the old chronometer methods and won't listen to the position line and intercept method at all—he actually says he doesn't think it is so reliable as the chronometer sight—Ye Gods!

'Man alive, it is wonderful how thoroughly stagnant some people become through sheer petty mindedness. The other day he sprung a wonderful 'new' method of Ex-meridian on me, more or less to take a feather out of my cap as we had been having a discussion on the navigation subject which arose from my extolling the prowess of aerial navigators who had never been to sea to learn navigation, and it hurt him badly. However, he was pleased to call his method 'Jeans's', the said gent having been an erstwhile teacher of navigation in the Navy, perhaps in the early 1840s. My popularity further decreased when I pointed out that it was really a direct method of spherical trigonometry and nothing very wonderful about it. This proved the last straw and we parted brass rags, the 'Old Man' making some trite remark about not believing in new-fangled ideas. It's hardly credible, is it?

'I've never been over the ship's rail since we left home saving money, oh yes! Good God, it takes a whole month's money to pay the rent.

'Do you think you could manage to get me the loan of a Royal Aircraft Establishment [bubble] sextant for practical sea tests? I'm awfully keen to try one at sea and I'm sure the results would be interesting, especially in foggy weather. You see, there are lots of days when the sun is quite bright and the horizon awfully hazed, and in the Red Sea during the day the clearest horizon is useless owing to the abnormal refraction.

'I've had some very good sights this trip: even got a fix with the Sun, Moon and Venus one afternoon.'

After taking his discharge from *Ixia*, Ernest pursued a pioneering career in civil aviation with no little distinction until his death in the crash of the airship R101 in 1930.

Among his most notable achievements was the creation of a new Livery Company of the City of London, the Guild of Air Pilots and Air Navigators, modelled on the Honourable Company of Master Mariners.

Epilogue

SOME EIGHTEEN MONTHS AFTER ERNEST JOHNSTON WAS killed, his widow Daisy died.

Their surviving son Ernest Alfred, invariably known from his cradle as 'John' was thereupon taken into the home and family of Mrs Jess Scott, the young widow of Ernest's chief, Major G.H. Scott. Every summer Jess took her children to stay with her parents on Garelochside, and it was through them that John was introduced to Mr and Mrs George Linck and their two lovely daughters.

George Linck was none other but the youngest son of Captain Friedrich Linck who figures so prominently in the tale of the Ship Masters. His wife Frances, moreover, was the grand-daughter of Captain Frank Harvey.

In 1940 John Johnston was accepted in marriage by Isobelle, the younger Linck daughter, and in due course they produced their own two talented daughters, Rosalind and Caroline.

The link between the lives of all those splendid Ship Masters —the Johnstons, the Flemings, Livingstones, Browns, Watsons, Harveys and Friedrich Linck is now established.

It is Rosalind and Caroline.

Appendices

Appendix 1

THE FIFERS

ROBERT **JOHNSTON**
Aberdour c1630-1690
|
JAMES
1664-1699
Mariner
Aberdour
|
JOHN
b. 1688
Aberdour
Ship Master
|
GEORGE
b. 1717
Ship Master
Kirkcaldy
|
JAMES (HAMISH)
b. 1765
Ship Master
Inverkeithing

JAMES **BROWN**
North Queensferry
Ship Master d. 1699
|
JOHN
b. c1680 N. Queensferry
Ship Master
|
WILLIAM
b.1710
Ship Master
Inverkeithing
|
GEORGE
b. 1738
Ship Master
Inverkeithing
|
ELIZABETH
b.1767
North Queensferry

MARGARET
married
Andrew **GIBSON**
1760

JAMES (HAMISH) married 1787 ELIZABETH
|
WILLIAM
b. 1795 N. Queensferry
Ship Master

continued in Appendix 5

Appendix 2

THE GIBSON CONNECTION

? GIBSON
Dunfermline

(Old) ANDREW
b. c1710
Carrier
Dunfermline

WILLIAM
b. c1730
Carrier
Dunfermline

ANDREW
b. 1767

ANDREW
b. c1740
Entrepreneur
married 1766
Margaret **BROWN**

ROBERT
b. c1712
Entrepreneur

ROBERT
b. c1739

JAMES
b. c1790
Owned Brig
Mary

WILLIAM LIVINGSTONE married ANN **GIBSON**
Ship Master 1759
North Shields

continued in Appendix 4

Appendix 3

THE FLEMINGS

WILLIAM **FLEMING**
1727-1780
Boatman
Sunderland

JOHN	GEORGE	HEDWORTH
b.1756	1764-1822	b. 1758
Seaman/Labourer	Ship Master	Ship's Carpenter

Children of JOHN:

JOSEPH	JOHN	DANIEL
1806-1859	b. 1797	b. 1805
Ship Master	Ship Master	Ship Master
Monkwearmouth		

Son of JOSEPH:

JOHN
1828-1883
Ship Master
Tynemouth

Daughter of JOHN:

ISABELLA LILLIE b. 1857 married JAMES JOHN **JOHNSTON** 1853-1903, Ship Master, North Shields

See Appendix 5

Appendix 4

THE LIVINGSTONES

WILLIAM **LIVINGSTONE**
b. c1730
Mariner, North Shields
Married Anne **GIBSON**
|
JOHN
b. 1760
Mariner
North Shields
|
WILLIAM
1784-1873
Soldier & farm worker
|
FRANCES
1819-1893
married JAMES **JOHNSTON**

See Appendix 5

Appendix 5

THE TYNEMOUTH JOHNSTONS

WILLIAM **JOHNSTON**
b. 1795 North Queensferry
Ship Master
Tynemouth

|

JAMES
1819-1872
Ship Master
North Shields
married FRANCES **LIVINGSTONE**

|

JAMES JOHN
1853-1903
Ship Master
North Shields
married ISABELLA LILLIE **FLEMING**

|

ERNEST LIVINGSTONE
1891-1930
Mariner & Aviator

|

The Author
married 1940
ISOBELLE HARVEY **LINCK**

|

ROSALIND CAROLINE

Appendix 6

THE HARVEYS & WATSONS

WILLIAM **HARVEY**	JAMES **WATSON**
fl. 1750-60	b. c1710
Ship Master	Ship Master
Port Glasgow	Port Glasgow
|	|
JAMES	WILLIAM
b. c1745	b. c1760
Ship Master	Mariner
Port Glasgow & Alloa	Port Glasgow
|	|
JOHN	WILLIAM
1774-1835	1790-1843
Ship Master	Accountant
Alloa	Greenock

FRANCIS ERSKINE (Frank) married MARGARET
1815-1863 — b. 1817
Ship Master
Alloa & Greenock

WILLIAM b. 1815 Ship Master
GEORGE b. 1821 Ship Master
ROBERT b. 1826 Ship Master

CHARLOTTE ERSKINE
|
FRANCES ERSKINE married GEORGE EMIL **LINCK**
son of Friedrich Linck
|
ISOBELLE HARVEY
|
ROSALIND CAROLINE

Acknowledgements

I am grateful to my publisher, Mr George Mann, for his unwavering and friendly support for this project and for the meticulous attention which my dear wife Isobelle has given to reading the proofs of this, as of my other books.

I must acknowledge the generous help of many correspondents, not least those related to me however distantly, and in particular Mrs May Thomas; Mrs Vera Houlbrook; Mrs Ida Rivers; Mrs Anne Livingstone; Mrs Michael Roberts; Dr Rita Hutcheson; Mr Charles P. Neat; Mr James A. Thompson; Mr L.P. Jarman; Mr Michael Scott, Keeper of the Christ Church Parish Registers at North Shields; The Rev. John Hucker, Minister of St Columba's United Reformed Church, North Shields; Mr John Boon, Keeper of the Records of the Scotch Church in North Shields; Mr Nicholas Robinson, Director of Stag Line Ltd; Mr Nigel Yates, Archivist at Tynemouth Central Library, North Shields; Miss A.E. White, Library Assistant at the National Maritime Museum; Mr J.K. Mealor, Senior Librarian at Falmouth Branch, Cornwall County Library; Mr W.A.L Seaman, Durham County Archivist; Mr James K. Sharp, Librarian, Central Library, Dunfermline; Mr T.M. Dinan, Assistant Shipping Editor, Lloyds of London; Mr G.T. Knight, Cornwall County Council Local History Librarian, and Miss P.M. Baxendine, Director of the Scots Ancestry Research Society. I hope that those whom I have omitted from this list will accept an old man's apologies for the scrappiness of his records, the illegibility of his notes and the failings of his memory.

As for sources, the gathering of the material for this book has taken many years of patient research in England and Scotland. First among my sources have to be the tales my Grandmother, Isabella Johnston, told me long ago; the second those which my wife's twin Aunts May and Belle Wilson told; the third the collection of ancient correspondence inherited from them relating to the Watsons and Leckies; and fourth, the Harvey Family Bible. Some old gravestones, notably in Scotland where I received useful guidance from Parish Memorial records in the Library of the General Register Office, have provided useful information. The Librarians of countless branches of County Libraries throughout the Kingdom have invariably been helpful.

My principal sources, however, have been the old Scottish Parochial Records, the Census Returns, and the Indexes of Wills in the Scottish Register Office, Edinburgh; the Shipping Records, Customs Records and Kirk Session Records in the Scottish Record Office; the English Registers of Births Marriages and Deaths; the Nonconformist Registers of England including Scottish Churches in England; the Census Returns; The Statistical Account of Scotland 21 Vols (1791-9); The New Statistical Account 15 Vols (1845); the Scottish Library, Edinburgh; the Mitchell Library, St Andrews; the Library of the Society of Genealogists, London; Registers of Shipping at sundry individual Custom Houses; the Board of Trade Registers of Masters and Mates, Muster Rolls of the Northern Ports and sundry other shipping records in the Public Record Office at Kew.

The historical department of Lloyds Register of Shipping has been helpful in tracing the movements of many vessels.

As might be expected of an island people, there is a vast literature of seafaring, of which the following selection has been particularly useful in providing background detail:

Alton and Holland *The King's Customs* 2 vols 1910
Bass, G.F. (Ed) *A History of Seafaring* 1972
Ballingall, Wm. *Shores of Fife* 1872
Bedford, Admiral Sir F.G.D. *The Sailor's Handbook* 1906
Chapelle, H. *The Search for Speed under Sail* 1700-1855 1968
Childers, Spencer *A Mariner of England* 1908
Conrad, Joseph *The Mirror of the Sea* 1906
Cook, E.W. *Shipping and Craft* 1829 (Facsimile pub. 1970)
Course, Capt. A.G. *The Merchant Navy, a Social History* 1963
Dana R.H. *Two Years before the Mast* 1840
Dana, R.H. *The Seaman's Friend* 1845
Davis, Ralph *The Rise of the English Shipping Industry In the 17th and 18th Centuries* 1962
Deacon, G.E.R. (Ed) *Oceans* 1962
Finch, R. *Coal from Newcastle* 1973
Finch, R. *The Ship Painters* 1975
Finden, W. & E. *Ports and Harbours of Great Britain* 1838
Flinn, M.W. *The History of the British Coal Industry 1700-1830* 1984
Gower R.H. *Seamanship* 1808
Hall, Capt. Basil, RN, FRS, *Fragments of Voyages and Travels* 1850

Hassan, J.A. *The Supply of Coal to Edinburgh* 1972
Hydrographer of the Navy *Ocean Passages for the World* 1973
Lubbock, Basil *Last of the Windjammers* 2 Vols 1929
MacGregor, D. *Merchant Sailing Ships* 3 Vols
Mathies and Pearsall *Shipping - A Survey of Historical Records* 1971
Nicholls A.E., FRAS, *Seamanship* 1917
Smith, R. *Sea Coal for London* 1961
Stephen, W. *History of Inverkeithing and Rosyth* 1921
Thornton R.H. *British Shipping* 1945
Villiers, A.J. *The Way of a Ship* 1954
Whymper F. *The Sea* 2 Vols 1877
Willan T.S. *The English Coasting Trade 1600-1750* 1938
Williams N. *Contraband Cargoes, Seven Centuries of Smuggling* 1959